What people are saying about *Inspired Educator, In*

"For years, Jennifer Stanchfield has shown an avid i
applies to what teachers and learners do together. H
neuroscience, and education to achieve a synthesis that will intrigue many educators." **—Sam Wang, Associate Professor of Neuroscience, Princeton University**

"*Inspired Educator, Inspired Learner* is inspiring! A helpful synthesis of some of the most important principles, trends, and practices education now needs to help all students be successful in the 21st century." **—Bernie Trilling, 21st Century Learning Advisor, co-author of *21st Century Skills: Learning for Life in Our Times***

"Aside from being an amazing resource for meaningful games and classroom activities, Jennifer's work synthesizes so many of the best practices I've come across as a teacher. It reminds me of why those things work, and it helps me to see the link between my own practices and the amazing work I've seen from other educators and facilitators. Whether you need an idea for tomorrow, some philosophical food for thought, or some research for your next unit plans, *Inspired Educator, Inspired Learner* will help. This is an exciting book for an educator." **—Neil Poynter, Teacher/Advisor at The NET Charter High School, New Orleans, Louisiana**

"My most engaging and powerful lessons and activities come from Jennifer Stanchfield. When I reflect on what works best in my classroom, every lesson on the top of my list has an element of her wisdom. Now, all that powerful wisdom is in ONE book. Since I can't have Jen in my classroom at all times, this book is the next best thing! This is a must-read for emerging AND veteran teachers who want to empower students' creativity and deepen engagement and understanding in the classroom." **—Heidi Pauer, 2003 New Hampshire English Teacher of the Year**

"A concise merging and melding of neuroscience, philosophy of education, practical facilitation skills, and experiential activities. Jennifer has brought the varied components of experiential learning into clear focus in an easy to use format. Designed to guide any educator or trainer toward a positive learning outcome with most any demographic. The science is approachable, the skills are reliable, and the activities are doable." **—Gary Swyers, Senior Program Manager, The Soderquist Center for Leadership and Organizational Development, Siloam Springs, Arkansas**

"This book is a great resource for educators—a 'go-to' handbook that is extremely readable and user-friendly. The research cited gives it a strong backbone. It's not a 'pie in the sky' book, but one that actually backs up the ideas with specific strategies to improve our teaching. That's the kind of book teachers can use." **—Colleen Barrett, Middle School Teacher at Crossett Brook Middle School in Waterbury, Vermont**

"Keep this book within arm's reach. Jennifer Stanchfield's ability to connect the dots between the tenets of experiential education and brain-based learning are extremely validating to the project-based learning we do in our Academy for Global Studies. Innovative educational models coupled with useful activities make *Inspired Educator, Inspired Learner* an amazing resource for all educators. I imagine it will become the dog-eared, go-to resource for novice and experienced teachers alike and will remain relevant throughout the many waves of educational trends." —**Nicole Griffith, Director of the Academy for Global Studies at Austin High School, Austin, Texas**

"Jennifer Stanchfield successfully takes critical, yet complex concepts like social-emotional learning and 21st Century skill development and makes them practical, relevant and approachable by framing them through an experiential learning lens. The book offers an array of useful activities that educators, youth workers and trainers can use to effectively support learning. Her experience working with both youth and educators brings the concepts to life and provides readers with the confidence to implement these strategies themselves." —**Kara Bixby, Manager of Research & Evaluation, Saint Paul Public Schools Foundation**

"This books provides a valuable link in connecting research on brain-based learning to the powerful work of experiential educators. Science is catching up to what industry professionals have known for a long time—what we do works! This new body of work is proof of what Dewey and Hahn interjected into their pedagogues long ago. Evaluation and research in our field will continue to broaden and hopefully inform how this book will be of value while being utilized in 'traditional classrooms' and many other settings. I think Jennifer is a frontrunner and leader in our field, pushing the conversation on what it means to be a well-rounded and excellent practitioner of experiential education." —**John Lee, Director of Operations and Programs, Omaha Outward Bound School**

"I love this book! Jen not only provides experiential activities for students that motivate and engage but also explains why these activities are so important. She shows how scientific research and neuroscience now prove what great educational thinkers like John Dewey have been saying all along. These are tried and true activities that work in the classroom, they build community and increase student motivation and learning." —**Amy Kahofer Dalsimer, MEd, 1st & 2nd Grade Teacher, Thatcher Brook Primary School, Waterbury, Vermont**

"This book is a must have resource to engage learners and bring joy to teaching. Once you incorporate Jen's facilitation techniques, you'll never go back. I am excited to have a tool that succinctly summarizes so many best practices. This book provides the recipe to enhance learning experience for students of all ages and unveils the neuroscience behind why these strategies are so effective at improving the learning experience. It should be required reading for educators and applies equally well to school and out-of-school learning environments." —**Katie Sidorsky, Program Manager, Techbridge: Science, Technology and Engineering Program for Girls, Oakland, California**

inspired educator inspired learner

This book is dedicated to Paul Stanchfield for his endless support and encouragement
and to Janet Stoddard and Jasper Hunt,
two of the best role models for experiential, student-centered teaching.

Inspired EDUCATOR Inspired LEARNER

Experiential, Brain-Based Activities and Strategies to Engage, Motivate, Build Community, and Create Lasting Lessons

Jennifer Stanchfield

Published by:

Wood N Barnes Publishing
2108 N. Willow, Suite A, Bethany, OK 73008
(405) 942-6812

Photos and Illustrations:
Colleen Barrett, 125; Carrie Blake, 103; Valerie Bodet LOOP New Orleans, 45, 46, 208; Abby Brown's Classroom, 59; Chris Brown, 128; Ginny Cunningham, 151b; Marci Charles Team BRIDGES Memphis, 8, 51, 68, 129, 158, 159, 195, 206; Denisimo Photography for the Mentoring Partnership of Long Island Back Cover Author Photo, 13, 55, 90, 143, 204; Kimberly Dessureault School's Out Program South Burlington, VT, 28; John DiGeorge, 77; Yolanda D'Allesio, 120; Josh Farr, 64; Luong Huynh Vermont Youth Development Corps, 56; Antje Kastner Studio, 253; Erin Klein Kleinspiration, 58c; Matt May LOOP New Orleans, 215; United States Dept. of Health and Human Services Courtesy of Wikimedia Commons, 5; Denise Sather, 145; Jennifer Stanchfield, xi, xii, 1, 7, 9, 10, 11, 14, 15, 20, 21, 23, 25, 26, 31, 32, 33, 34, 35, 37, 38, 39, 41, 42, 43, 44, 52, 53, 58a and b, 60, 61, 63, 65, 69, 71 ,72, 73, 74, 75, 75, 78, 80, 81, 82, 83, 85, 87, 92, 93, 95, 97, 99, 102, 104, 105, 106, 109, 111, 113, 114, 116, 117, 119, 121, 123, 125a, 127, 131, 135, 140, 141, 148, 150, 151a, 152, 153, 154, 155, 156, 157, 160, 161, 162, 164, 166, 168, 169, 170, 171, 173, 175, 176, 177, 180, 183, 185, 186, 187, 190, 192, 194, 197, 198, 200, 207, 210, 213, 217, 221;
Paul Stanchfield, 178; Will Wise, Shaver's Creek Environmental Education and Team Development Center, 4b, 49, 89, 199.

Cover Photos by Valerie Bodet, Marci Charles, Denisimo Photography, Yolanda D'Allesio, and Jennifer Stanchfield

Cover Design by Blue Designs
Copyediting & Layout Design by Ramona Cunningham

Printed in the United States of America
Bethany, Oklahoma
ISBN # 978-1-939019-13-4

Contents

Introduction

Connecting the Dots Between Experiential Education, Brain-Based Learning, Differentiated Instruction, Social-Emotional Learning, and 21st Century Skills

Snapshot: There is laughter floating out the door of this eighth-grade classroom. All of the desks have been pushed aside, and the students are in a circle, standing on colored spot markers. They intently listen to one student who reads a question/prompt to the rest of the group: "Anyone who knows what the Transcontinental Railroad was?" Immediately, ten students rush to move from their spots in the circle to find another. The student who lands on the red spot enthusiastically shares his answer: "It was the first transcontinental railroad completed when the Union Pacific and Central Pacific railroads joined their tracks." Other students eagerly contribute: "It was built in 1869 and connected in Utah," and "It is important because it increased western settlement, because the journey became so much shorter and easier than it had by wagon." A student asks: "Is that why so many people left Vermont?" Another student asks, "How come we don't use railroads much anymore?" The teacher is pleasantly surprised as he observes these students who rarely raise their hands in class now actively engaged in a discussion about the early 1800s Westward Expansion era in the U.S.

Snapshot: The college professor has switched off her PowerPoint® following a 20-minute lecture. She has students gathered in a formation of two concentric circles at the front of the lecture hall. Each pair of students is actively discussing their reactions to the professor's lecture on land use, environmental impact studies, and recent legislation in the state of Oregon. The inside circle takes the perspective of a development company wanting to build a destination resort near the Metolius River in Oregon, while the outside circle represents the Confederated Tribes of Warm Springs who oppose this project. Students discuss the issues based on the facts presented during the lecture and their position in the concentric circles. After a few minutes, the professor asks the inside circle to move three people to the right and poses another reflective question to the group.

Snapshot: Second-grade students are lining up for lunch. An argument ensues at the front of the line about whose turn it is to lead the class to the cafeteria. A little voice from the back of the line says, "Hey, you two need to compromise about this like we did today during the game we played. We all just want to get to lunch." The girls look at each other and the rest of the line and then quickly decide who leads today and who leads tomorrow without any intervention from the teacher.

Snapshot: A group of hospital staff gathers at 8:00 a.m. for a day-long training on risk management policies and procedures. The HR staff person running the training is greeting each of them at the door as they enter. She has laid out pin-back buttons with humorous sayings (i.e., "Dangerous When Bored", "As Is", "Queen of Crisis") on a table by the coffee and bagels. She asks the participants to choose a button that reflects their mood or attitude this morning. The participants, who at first were reluctant to attend, begin to chat with each other about their buttons, setting a positive tone and increasing engagement as the workshop begins.

Connecting the Dots

A teacher affects eternity; they can never tell where their influence stops. —**Henry Brooks Adams, historian, educator, and novelist**

You might read the snapshots and think they are examples of differentiated instruction, social-emotional learning, responsive teaching, 21st century skills, and/or positive behavioral supports. In my mind, with all of the right elements in place, these snapshots are just examples of plain, good teaching regardless of what you label them. As an educator or group facilitator, you have likely studied or embraced many models and theories of teaching and training yourself (or at least been asked to explore them in a professional development in-service). You can probably list a number of programs with varied acronyms that have passed over your desk throughout the years.

Educators are often overwhelmed by the number of new initiatives being put forward by their schools or organizations. Some have become skeptical about what they refer to as the passing "fads" or "magic bullet" models or curriculum that arise in the field of education. Many schools or organizations repeatedly take on new methods and models of best teaching practices, then quickly move on to others before allowing the test of time and a true commitment to the program or methods. I too have seen this "flavor of the month" approach, but I also have seen effective educators pull valuable information from these models and theories. As good consumers of knowledge, these educators look to the evidence-informed practices that will work for them and their learners and put them to use.

With 25 years in the field of education and more than 40 years as a learner, I find time and time again that the fundamentals of the many models and approaches that work well can be described as examples of experiential education. The best of these approaches (i.e., the ones that inspire educators and engage learners) exemplify experiential education principles and what educational neuroscientists are now identifying as brain-based teaching approaches. **By connecting the dots between experiential education theory and methods, brain-based research, differentiation or personalized instruction, social-emotional learning, 21st century career readiness, and a strength-based attitude toward teaching, a successful and rewarding holistic approach to teaching and learning can be created.**

A century ago, educational philosophers and scientists such as John Dewey, Alfred North Whitehead, Lev Vygotsky, and Jean Piaget put forward the idea that learning should be more than just imparting knowledge. These philosophers promoted what I think of as a commonsense approach, emphasizing that successful teaching and learning require meaningful experiences and interaction with others in an environment that intentionally encourages collaboration, problem solving, inquiry, and reflection. They believed effective teaching influences the whole person—emotionally, physically, socially, and intellectually—and that educators should engage learners in relevant experiences that relate and connect to real life, preparing them to be active members in a democratic society. These ideas put forward nearly a century ago are now being supported by scientific studies of the brain and how people learn, retain, and apply information.

The emerging field of educational neuroscience or brain-based learning supports the experiential approach to teaching. This new field combines the best research coming from educational psychology, pedagogy, and neuroscience and offers useful information to help enhance our practice as educators (Fischer, Goswami, & Geake, 2010; Goswami, 2004; Sousa, 2010). The education field is now looking to the needs of today's society and integrating social and emotional skills such as critical thinking, problem-solving, empathy, and communication as essential parts of academic success and career readiness.

This aim of this book is to bring these enduring ideas and the research that supports them together, offering practical and meaningful ways to engage learners and create lasting lessons. The heart of this book is to bring joy to learning and teaching; educators will find ways to build respect and compassion for learners and increase a sense of empowerment, belonging and fulfillment in learning.

About This Book

My own experiences as an educator have included teaching and mentoring in Pre K-12 classrooms; serving as a recreational therapist, teacher, and group leader in child, adolescent, and adult psychiatric care facilities and community organizations; and providing professional trainings and courses for adult learners. These experiential programs have ranged from community gardens to recreation programs for inner city and rural youth to adventure education programs and trip-leading with participants of all ages. Though a focus of my work now is on training and providing workshops for teachers, counselors, youth workers and other educators, I still spend a considerable amount of time working directly with students in elementary, middle, high school, and college classrooms. The practical methods and techniques offered in this book come from my ongoing experiences in classroom and group facilitation settings. They are supported by a convergence of evidence from philosophy, effective practice, and multiple fields of research, including neuroscience, pedagogy, and educational psychology.

This book is designed to inspire educators with activities and strategies they can apply immediately to their work in the classroom or group room to create lasting lessons. The goal is to increase engagement and motivation by developing a positive, collaborative learning community. The following topics are addressed and expanded upon throughout:

- Experiential, brain-based methods that engage learners of all ages emotionally, physically, intellectually, and socially
- Engaging, community-building activities to develop rapport, respect, positive communication, collaboration, and creativity
- New research from the field of educational neuroscience and its impact on learning and social and emotional skills development
- Interactive games and activities to enliven academic or training curriculum and differentiate instruction

- Methods for increasing involvement, buy in, and ownership in learning experiences
- Practical ways to develop a healthy climate of respect and trust, peer-to-peer and educator-to-student
- A variety of brain-based teaching approaches and activities that will motivate learners and promote critical thinking, creativity, and collaboration
- Experiential techniques for building positive peer relationships, developing character, and practicing social-emotional skills such as problem-solving, empathy, communication, and conflict resolution skills
- Active, brain-based strategies to facilitate meaningful reflection and dialogue for review, content delivery, and formative assessment
- Participant-centered learning techniques to encourage learners to take more control and responsibility for their learning
- Innovative reflective tools and techniques to increase relevancy, meaning, and connection to future learning situations

These methods can inspire your creativity, add to your teaching and group facilitation toolbox, and encourage new perspectives on experiential learning and differentiated instruction. The strategies you take away can empower you to build a supportive learning community and create lasting lessons, helping participants develop the important skills they will need to be productive citizens and life-long learners.

> Learning promotes learning. Engaging in the process of learning increases one's capacity to learn. —**Judy Willis, neurologist & teacher, 2008**

A Map of This Book

This book is a combination of theory and practice, art and science. Throughout, I share my favorite "tried and true" activities and methods. The activities are blended with theoretical and brain-based concepts. Stories from my experiences are included to illustrate the point. The "why" behind these methods and activities is explored, referring to the brain research and pedagogical models in support of these approaches that help us become better evidence-informed educators.

Strategies and activities are indexed and cross-referenced. A great deal of effort is made to connect the social-emotional, active review, formative assessment, problem-solving, community-building, reflective and academic applications of activities and strategies. Although each chapter builds upon the next, a reader can dip into the book at any point to take away useful tools and techniques, it doesn't have to be read from front to back.

Note: Most of the activities and strategies in this book can be adapted to fit a broad range of ages and abilities. I include **Purpose/Focus, Materials** (where applicable), **Facilitation Suggestions**

and other pertinent information with each activity in the book. However, I purposely did not categorize activities by specific age ranges or types of audiences, as I believe most activities can be adapted to fit a wide range of participants and settings. If a particular activity or strategy doesn't fit your specific group, it might inspire you to think of a similar approach that will.

Chapter One: It's an Exciting Time to Be an Educator builds a scientific and philosophical foundation for the activities and strategies offered throughout this book. It explains how the emerging field of educational neuroscience or brain-based learning and teaching supports experiential education philosophy and methods. Exciting information about research on the brain and learning is offered, including neuroplasticity, multiple pathways to learning, the impact of emotions on learning, executive functions, joyful learning, and attention and retention. This chapter underscores the value of physical activity, social-engagement, play, and meaningful reflection for improving learning outcomes. The activities and strategies offered in subsequent chapters are based on the philosophical and scientific underpinnings explored in this chapter.

Chapter Two: Innovative Approaches to Teaching offers an interconnected framework for facilitating experiential, brain-based teaching that includes the models of differentiated instruction, social-emotional learning, 21st century skills, and a strength-based approach to teaching and group development. The commonsense aspect of these approaches is emphasized, encouraging educators to look at the fundamental, interconnected principles of the models that can help us improve our practice. Using the best of these methods, we have an evidence-informed, strength-based approach to teaching that is sustainable and effective—not a passing fad.

Chapter Three: The Power of Play and Collaborative Learning explores the rationale for using interactive activities, games, and problem-solving challenges to teach both academic and social-emotional skills. This chapter underscores the benefits of a cooperative, playful approach to skills-building and offers information from the field of educational neuroscience and pedagogy to support these methods. We also touch upon the history and practical applications of using play and collaboration as a teaching modality for learners of all ages.

Chapter Four: The Inspired Educator offers practical suggestions for facilitating experiential activities and for creating ownership and buy-in on the part of learners. We discuss the importance of understanding "the why" behind choosing an activity and the art and science of sequencing and scaffolding lessons. The concepts of a learner-centered approach, the educator as guide, the importance of attitude, collaborative teaching experiences, and choice and control in learning are explored. Practical suggestions are offered for re-imagining the learning space. Facilitation techniques include attention getters, ways to form groups, building positive group norms, engaging reluctant group members, and cultivating creativity.

Chapter Five: Strong Beginnings offers practical, brain-based strategies to "hook" or engage learners from the very beginning of a lesson or experience. These hooks create context for lessons, draw learners into experiences, build a positive and supportive learning environment, and spark dialogue, reflection and inquiry.

Chapter Six: Get Them Moving, Talking, Reflecting and Keep Them Engaged provides active and social strategies to engage learners in reflective dialogue, academic content review, and formative assessment. These methods cultivate multiple pathways to learning and increase engagement, attention and retention. This chapter demonstrates how many activities traditionally used as icebreakers at the beginning of a program to facilitate conversations and interpersonal connections can be re-purposed with an academic focus, integrating these multi-modal approaches with direct instruction and ongoing formative assessment.

Chapter Seven: Making the Most of Your Time With Multi-Purpose Activities That Teach picks up on the theme of active engagement offered in chapter five and adds options for using activities that integrate academic and training content with group building, problem-solving and social-emotional skills development. These activities can be used to reinforce academic content and provide formative assessment. They will inspire your thinking about experiential, brain-based methods you can use to increase academic outcomes while promoting interpersonal and life skills.

Chapter Eight: Building a Strong Foundation for Learning and Life offers specific activities and lesson ideas for building a supportive learning community and developing the social-emotional and 21st century life skills that learners need to thrive in modern day society. These practical approaches and lesson planning ideas build rapport, healthy trust, and positive group norms. They are useful for educators (i.e., trainers, facilitators, teachers, principals, college student life program staff, youth workers, adult educators) looking for engagement and group-building strategies.

Chapter Nine: Bringing Learning to Life With Reflection helps learners make connections between educational experiences and real life situations. Many educators recognize the value of reflection, but find that facilitating it is one of the most challenging aspects of their work. This chapter offers engaging reflective methods to increase depth of understanding and help learners apply lessons to real life and future learning.

Chapter Ten: The Reflective Educator emphasizes the importance of prioritizing time for self-reflection to improve your practice as an educator. Reflecting on one's professional practice increases a sense of meaning and purpose around one's work and offers insight into what strategies or approaches are most effective. This ongoing reflection helps educators use what they learn each day from students or clients to improve their work in the future.

How to Use This Book

As mentioned above, many educators might use this book as they would a recipe book, flipping through the activity sections to find the activity or teaching tip that fits a specific goal or lesson topic for their group. Others might find it useful to start with the beginning chapters that explore the "why" behind what we do in experiential education. Whether you are a veteran or novice educator these chapters might give you some new perspectives on the connection between the enduring philosophy of experiential education and the modern day research that supports it. Teachers can reference this information to explain why their students are away from their desks and playing games as part of academic time. A professional trainer could use it to explain why they put the PowerPoint® away and arranged chairs in a circle.

This book was designed for readers to delve in at any point to find what they need. Skip around; come back to chapters one or two when you want some theory and science to support what you do. Hit chapter nine when you are looking for meaningful reflective tools and techniques, or come back to chapter four when you want some facilitation tips and tools to smooth out new activities. No matter where you start, you will find tools, techniques, and insights to inspire your practice as an educator or facilitator.

chapter 1

It's an Exciting Time to Be an Educator

Evidence-Informed Practices That Inspire Educators and Engage Learners

Snapshot: Around the room you see model bridges made to scale out of spaghetti and straws. Groups of students are explaining to attendees how they chose their bridge design, met (or didn't meet) budget and construction requirements, and what role they took in the process. For the past five weeks, the students have worked together, taking on the real life role of architect, engineer, project manager, or accountant. Students were challenged to find ways to communicate, resolve conflict, and create positive group norms. Binders are on display, showcasing the extensive work accomplished during this cross-curricular project. Tension builds as the math and science teachers move about the room, visiting each bridge with a set of weights to test load-bearing capacity. Onlookers hold their breath each time a bridge is tested.

What is especially meaningful for those gathered here is their renewed understanding of the importance of bridges and bridge builders to their everyday life. A recent hurricane destroyed a number of bridges throughout their town leaving many people stranded. They had observed the reconstruction of these bridges first hand and become friendly with the state engineers and bridge construction companies in the area. These professionals took the time to visit the school halfway through the project for a panel discussion. They were so impressed with the students and this project that they drove for two hours to attend tonight's presentation.

Enduring Themes and Supporting Science

The most important attitude that can be formed is the desire to go on learning.
–John Dewey, 1938

It is an exciting time to be an educator; especially an educator who believes in experiential approaches to teaching and asserts that an educator's role is to take into account the whole person, preparing him or her for an active and productive role in society. As educators, we have more information available to us than ever before about how people learn best and the conditions that enhance learning outcomes. Advances in neuroscience allow researchers to study the living brain and how people of all ages are motivated to learn, attend to, store, recall, and transfer information.

This research suggests the brain has the potential to change and develop more throughout childhood, adolescence, and adulthood than previously thought. Ongoing practice of reinforcement and reflection—tenets of experiential education—enhance skills, improve habits, and promote learning throughout life. As neurologist, teacher, and author Judy Willis (2013) states, "Educators have the important role of being 'caretakers' of learners' brains."

It is clear that the role of educator goes beyond the job of solely transmitting information. Educators need to prepare learners for a rapidly changing society and teach them to become innovators. Supporters of 21st century skills development argue that teaching methods need to be authentic, relevant, and collaborative. Social-emotional skills such as problem solving, communication, critical reflective thinking, creativity, character, and citizenship are needed to prepare people to thrive in and contribute to modern society. This is the same argument made a century ago by John Dewey and other educational innovators and proponents of experiential education such as Alfred North Whitehead (1929), Lev Vygotsky (1933), Carl Rogers (1969), and Jean Piaget (1926). The experiential view of teaching commissions the educator to act as a guide whose role is to facilitate, create conditions for meaningful learning experiences, and ignite a desire to pursue lifelong learning. Now we have a greater understanding of the biological basis behind why these enduring themes of experiential education work.

This chapter will inspire you with exciting new information about the brain and learning and why differentiation is worth the effort. For readers who haven't bought into experiential approaches, this chapter will provide a scientific basis to help you understand why experiential education philosophy and methods endure. For those who already consider yourself experiential educators, this information will help articulate your practice and explain its value to skeptical colleagues, administrators, or parents who worry these approaches mean time away from focusing on academic or training content. Most importantly, this chapter will empower you with exciting new information about the brain and learning to make informed choices in teaching and facilitation that will inspire and engage learners.

What is Experiential Education?

How is it relevant to my practice as an educator?

> *Give the pupils something to do, not something to learn; and the doing is of such a nature as to demand thinking; learning naturally results.* — **John Dewey, 1916**

The experiential approach to education and group work is based on the idea that optimal change and growth take place when people are actively (physically, socially, intellectually, emotionally) involved in their learning rather than just being receivers of information. As stated in the introduction, this approach emphasizes that learners are shaped by their experiences, and effective teaching involves meaningful experiences and interaction with others in an environment that intentionally encourages collaboration, problem solving, inquiry, and reflection.

John Dewey, a prolific twentieth-century educator, psychologist, and philosopher, is considered by many the "father" of experiential education. He promoted this philosophy of education, which he termed a philosophy of experience and education, or "experimental education"—later coined by others as experiential education. During the progressive movement in education, Dewey along with others aimed to move educational practices out of what many thinkers believed had become the narrow and limited realm of modern education. Dewey (1910, 1925, 1934) believed educators should make more of an effort to understand how learning takes place and take into account the nature of the human experience to better design lessons/learning experiences. He felt modern education was ignoring the commonsense observation dating back to early Western and Eastern philosophers (Plato, Socrates, and Lao Tzu among others). Dewey maintained people learn most when they are actively engaged with their whole selves, find the material meaningful and relevant to real life, and feel a sense of control over learning situations (Dewey, 1897, 1900, 1938; McDermott, 1981).

Dewey's philosophy also embraced reflection in learning (1916, 1925). A tenet of experiential education is that in order for learning to truly occur, learners should be provided with opportunities to reflect on their educational experiences so they relate and transfer to real life. He emphasized that schools should prepare students to be active members in a democratic society. In this approach, learners should be encouraged to experiment and think independently with support and guidance from educators (1900, 1910, 1938).

Though a learner-centered approach is key to this philosophy, Dewey did caution his followers and other proponents of experiential education against what he saw as a misrepresentation of his ideas: that learner-initiated, active engagement alone was the best mode (1902, 1938). Instead, Dewey advocated for a balanced and structured approach to teaching—taking into account the content delivery along with the interests and experiences of the learner. He believed in the importance of activity, freedom, and experience but differed from other progressive educators in his emphasis on the importance of structure, purpose, and reflection in learning experiences (1902, 1910, 1938; McDermott, 1981).

According to Dewey, the role of the educator was not to be the all-knowing transmitter of information to passive students, but rather a facilitator and guide. In his 1897 Pedagogic Creed, he stated: "The teacher is not in the school to impose certain ideas or to form certain habits in the child, but is there as a member of the community to select the influences which shall affect the child and to assist him in properly responding to these influences" (McDermott, 1981, p. 447).

These ideas put forward nearly a century ago are now supported by scientific studies of the brain and how people learn. In recent decades, many advances have been made in the neuroscience field. Brain-imaging, brain mapping, and other technologies have allowed neuroscientists to expand their studies of the living brain and identify optimal conditions for learning. These conditions include creating active, novel, and relevant learning experiences combined with meaningful reflection. Like Dewey, educational neuroscientists promote the importance of giving learners a sense of choice, control, and ownership in their learning. Studies of the brain are showing that physical, emotional, and social involvement in learning increases engagement and retention (Medina, 2008; Willis, 2006, 2010a). Like experiential educators, proponents of brain-based learning stress the importance of creating opportunities for student reflection and feedback from peers as well as teachers so that lessons can be applied to real life and future learning.

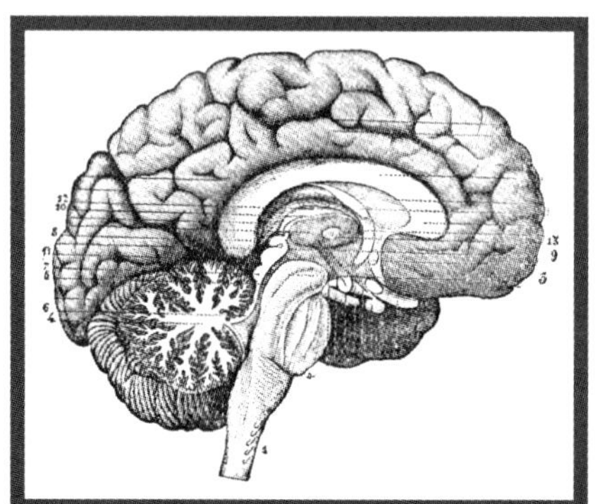

Experiential Teaching and Group Facilitation

The term "facilitator," as used in this book, describes the role of a person in the field of education, counseling, recreation, training, or other related areas that works to help individuals and groups create positive change, learn new skills, and gain new perspectives. Experiential facilitation is an intentional approach based on the idea that people learn and change more from the process of working through problems and finding solutions than from being given answers by a teacher/counselor/leader. Classroom teachers might find that thinking of themselves as a facilitator or guide in

facilitate: to make easier, aid, assist, smooth the progress of, to make possible, create, compose.

the process of learning and discovery is key to helping students create ownership of their learning, thus increasing their effectiveness and satisfaction as a teacher (Stanchfield, 2007).

A benefit of experiential facilitation is that, by using a variety of methods and combining action and reflection, educators can better differentiate their lessons to reach the varying needs and styles of group members or students with which they work. Experiential approaches to teaching and facilitation can enhance one's ability to engage and motivate learners, inspire a sense of discovery, instill a desire to learn, and create a positive and supportive learning community.

What is Brain-Based Learning?

How can information from the field of neuroscience impact my teaching?

> *There is no scientific study more vital to man than the study of his own brain. Our entire view of the universe depends on it.* —Francis H. C. Crick (from Scientific American, September, 1979)

The brain-based approach to teaching aims to help educators find the best research on teaching, learning, and the brain and apply that knowledge in their day-to-day practices to improve learning experiences and outcomes. In the past, there has been minimal focus on brain development, structure, and function as part of education training. Interaction between educators and neuroscientists has been rare.

During the past 20 years, especially the last ten, brain researchers have been able to study the living brain in ways they never have before, using MRI, brain mapping, and PET (Positron Emission Tomography) scans among other new technologies. They can view what happens in the brain when information enters and is categorized into short and long-term memory. Judy Willis (2013), a neurologist, teacher, and one of the leaders in this emerging field of educational neuroscience states: "Scans can literally show learning taking place."

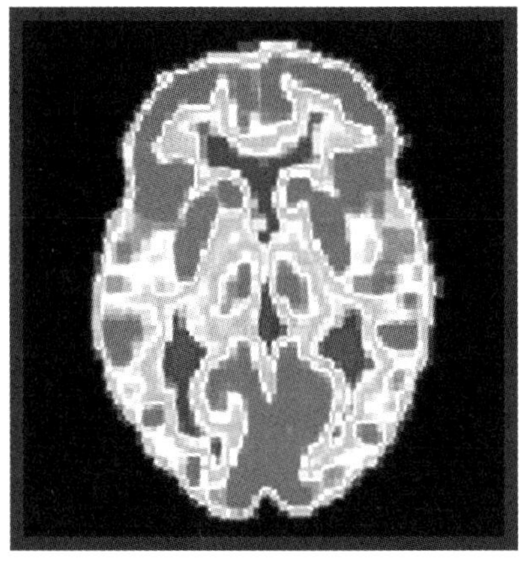

Brain-Based approaches to teaching and group facilitation can be defined as the intentional use of strategies that are based on recent research on the brain and learning.

Educational Neuroscience is known by many as the field of mind, brain, and education. This field pulls together scientific research from neuroscience, cognitive neuroscience, educational psychology, education theory, education technology and other fields to promote evidence informed teaching approaches. Proponents of this field believe that research on the brain and learning can be applied to education in a practical and meaningful way (Goswami, 2004).

Research has dramatically changed what we know about the brain's chemistry, structure, and function. Proponents of brain-based teaching and the emerging field of educational neuroscience believe it is imperative for educators to have an understanding of brain structure and function in order to improve their practice and help people understand how they learn best. There has been concern on the part of neuroscientists that information from brain research can be oversimplified or misrepresented in the educational field (Sousa, 2010). The growing educational neuroscience field aims to open the lines of communication between brain researchers and educators, bringing the best of quality research to front line educators (Learning & the Brain Society). This emerging field combines brain-research, educational psychology research, and other pedagogical research that suggests best practices for educators.

In Boston, at the 2013 Learning & the Brain Conference, experiential methods were referred to as a best practice in many workshops and keynote sessions. Though there is still much that is unknown about the brain, what has been learned supports and informs experiential education philosophy and methods. These findings from neuroscience are not just confirmatory, but explanatory because they can help educators make better choices about how to structure and facilitate learning experiences.

As an experiential educator, the methods I use to be more intentional and deliberate in designing my lessons are based on research on neuroplasticity, multiple pathways to learning, attention and retention, the impact of movement and emotions on learning, and the relationship between executive functioning and social-emotional learning. These findings from neuroscience give support to and align with the long-held philosophical tenets of experiential education and explain the importance of practices such as differentiation and social-emotional learning.

In this chapter, we explore the basics of these topics and their relevance to educators. This lays the foundation for the practical approaches used to enhance learning throughout this book. Table 1.1 on pages 16-17 is a summary of how experiential education principles connect with the brain-based tenets explored in this chapter.

Neuroplasticity: We Can Change Our Brains!

> *Learning changes your brain. When you learn something new, or form a new memory, new connections are made between neurons. The act of learning can actually create new dendrites and new connections.* —**Shelly Carson, *Your Creative Brain*, 2010.**

John Dewey and Jean Piaget both promoted the idea that people are shaped by their experiences, but the commonly held view of neuroscientists for most of the 20th century was that brain development was relatively complete by early childhood. This belief is being challenged by advances in neuroscience over recent decades. Brain researchers have found that many parts of the brain change and develop throughout life in response to experience, practice, and training (Aamodt & Wang, 2011; Willis, 2012). This concept called neuroplasticity describes how networks change in relation to their use. Repeated activation of a circuit leads to the addition of synapses and dendrites and the production of layers of insulating myelin

Practice is the best of all instructors.

— Publilius Syrus, first century BC Latin writer

around axons. This repeated activation through practice and experience leads to a strengthening of that pathway. When circuits are activated by experience and thinking about that experience, the electrical impulses stimulate more dendritic and synaptic connections, forming more layers of insulation (myelin). This increased insulation speeds up information transmission, processing, and memory retrieval (Willis). When pathways aren't used, synapses are pruned (they become weak and disappear). Like muscles, neural pathways are strengthened through use.

This evidence supports the tenets of experiential education that promote teaching practices involving positive, relevant learning experiences. These teaching experiences are best when combined with practice, guided reflection, and review. What is especially exciting about new research in plasticity is that not only does practice improve academic learning but that the areas of the brain appearing to have a great deal of plasticity are the areas that control executive functions (Diamond, 2010; Aamodt & Wang, 2011). These functions include social-emotional and 21st century life skills such as self-control, decision-making, reflection, critical thinking, and communication. In the following chapters, we explore a variety of strategies that engage learners in meaningful and relevant practice of academic and social-emotional skills.

Create Multiple Pathways to Learning

You don't understand anything until you learn it more than one way. — **Marvin Minsky, artificial intelligence pioneer**

The more senses that are used to learn and practice information, the more neural networks are activated and the more ways the brain is able to store and retrieve information. Neuroscientists emphasize that long-term memory storage requires spaced repetitions of input (Kandel, 2006 cited in Willis, 2013). Multiple strategies for learning and practicing information such as moving and talking, active reflection, art, and writing not only increase learners' interest and attention but also create multiple pathways to learning. Sam Wang states, "Memories are reinforced and even transferred to other brain regions than the place where they are stored for the first time. This movability of memory means that memory can change with time—and be strengthened with repetition. Like I tell my students: cells that fire together, wire together" (personal communication, April 19th, 2014).

The more ways something is learned and practiced, the easier it is to recall and access that information. This book offers a number of experiential strategies for engaging learners in ways that use multiple senses to learn, practice, and reflect upon information to increase multiple pathways to learning.

Emotions are Integral to Learning

> *The message from social and affective neuroscience is clear: No longer can we think of learning as separate from emotion.* —**Mary Helen Immordino-Yang and Matthias Feath, 2010**

Emotions influence the way the brain filters sensory input (Aamodt & Wang, 2011; Willis, 2010a). As mentioned previously, our brain has a complex information filtering system. Emotions draw our attention and impact what input is filtered through our senses and how we respond to them immediately and in the future. We remember emotionally arousing experiences but those that are emotionally stressful can negatively impact people's ability to learn. Our attitudes toward learning are built upon prior experiences. Our emotional responses trigger memories, recollection of information and even subconsciously affect how we approach a task or lesson (Immordino-Yang, 2012). John Dewey proposed a century ago that learners need to feel an emotional connection to learning in order to create lasting lessons, and modern day educational neuroscientists are finding in their research that this holds true (Immordino-Yang & Damasio, 2007; Immordino-Yang & Faeth, 2010). Mary Helen Immordino-Yang and Matthias Faeth suggest that educators intentionally foster emotional connection to the material being learned. Like Dewey, they suggest learner-centered choice and control increase this emotional connection to learning and promote student involvement in designing lessons. They emphasize the importance of relating lessons to the lives and interests of students, of solving open-ended problems, and reflecting on experiences to develop what they call skilled intuition. When these emotional connections are cultivated and reflected upon in meaningful ways they will transfer to other academic and real-life situations (see chapter nine on reflection).

> I've learned people will forget what you said, people will forget what you did, but people will never forget how you made them feel. — **Maya Angelou, poet, novelist, & Civil Rights activist**

Joyful Learning

Joy: The emotion of great delight or happiness caused by something good or satisfying.

When I think back to my own most powerful learning experiences, I realize joyful learning was taking place. There are moments that arise in my work when I see learners express delight in their accomplishment, share a deep connection to the material, or become so engaged in a lesson or project they lose all sense of time. When this occurs, joyful learning is taking place.

Psychologist Mihaly Csikszentmihalyi (1990) has repeatedly underscored the connection between joy, intrinsic motivation, high attention and productivity. Educational neuroscientist Judy Willis (2010a, 2013) promotes the importance of joyful learning, pointing out that high-stress instructional practices such as fear or punishment negatively impact our brains' ability to take in and synthesize information. Conversely, emotionally rewarding experiences increase engagement, attention, and retention. When we cultivate joyful learning—not just having fun but encouraging a sense of accomplishment, connection, pleasure in learning, positive social connections, and deep interest—we enhance learning outcomes.

Willis (2013) emphasizes the influence of dopamine on learning. Dopamine is a nuerotransmitter—a chemical released by nerve cells to send signals to other nerve cells. Pleasurable experiences such as play, positive feedback, and a sense of competency can increase dopamine. When dopamine is released during enjoyable experiences, it increases one's ability to attend to and store memories. Willis states, "Learning activities that can induce the release of dopamine and create these pleasurable experiences include physical movement, personal interest connections, social contact, novelty, intrinsic reward, a sense of achievement, choice, play, and humor" (2008, p. 7). In her presentations and articles she often suggests that educators look to the field of computer game design as a model for promoting engagement in teaching. She emphasizes that what makes computer games so attractive and addictive to many people is the incremental achievable challenge combined with ongoing regular feedback on progress toward a goal. This increases emotional engagement and dopamine and therefore intrinsic motivation. Willis believes educators can create this kind of engagement by incorporating these principles into teaching (2011a).

> What we learn with pleasure we never forget.
>
> — **Louis Sebastian Mercier, French writer, dramatist**

Get Their Attention and Keep It

A strong link between attention and learning has been shown in classroom research both a hundred years ago and as recently as last week. The story is consistent: Whether you are an eager preschooler or a bored-out-of-your-mind undergrad, better attention always equals better learning. It improves retention of reading material, accuracy and clarity in writing, math, science—every academic category that has ever been tested. —**John Medina, *Brain Rules*, p. 74**

In his book, *Brain Rules,* John Medina stated that in typical lecture presentations people tend to check out after 10 to 15 minutes of lecture. This statement intrigued me, and I started looking at more research on attention. Medina (2008) and Willis (2012) emphasized that we have so much data coming in through our senses—millions of sensory neurons are sending messages—only a few will grab our attention while the rest are ignored. The messages that get through are connected to emotions, novelty, and relevancy. We must find the information relevant, attractive, or important to the success of our survival. Dewey knew this, and brain researchers are finding those are the messages our limbic system "allows through." From the time we are babies, we select relevant information for attention and disregard most of the rest. Current neuroscience and pedagogy research validates the idea that educators will increase attention, motivation, and learning outcomes when they weave in opportunities to get learners away from their desks and interacting with their peers. In upcoming chapters, we will explore practical ways to increase attention, retention, and active engagement from the moment learners walk in the door.

Get Them Moving: Physical Activity Improves Learning

It may be seriously asserted that a chief cause for the remarkable achievements of Greek education was that it was never misled by false notions into an attempted separation of mind and body. —**John Dewey, 1916, p. 166**

There is a growing body of evidence demonstrating the positive impact of exercise on brain development and function—and therefore memory and learning throughout life. Exercise has the long-term effect of improving blood flow, thereby improving the flow of oxygen and nutrients to the brain. Neuroscientists have found exercise also increases blood flow in certain regions of the brain such as the hippocampus—an area of the brain that plays an important

Exercise is the single most powerful tool you have to optimize your brain function. — **John Ratey, *Spark***

role in the organization of memory and the transfer of information from short term memory to long-term memory (Chapman et al., 2013). Research has shown that exercise increases the survival of newly formed brain cells in the hippocampus of lab animals and both young and older adults. It also causes the release of growth factors that support the development of dendrites and synapses and synaptic plasticity in the brain—the ability of the brain to change (Aamodt & Wang, 2011). "Physical exercise may be one of the most beneficial and cost-effective therapies widely available to everyone to elevate memory performance" (Chapman, 2013).

Antronette Yancy (Toni) was known as a luminary in the public health field and regularly addressed the importance of physical activity to promote learning and work success for people of all ages. Yancy was a professor in the Department of Health Services, UCLA School of Public Health, and author of *Instant Recess: How to Build a Fit Nation 10 Minutes at a Time* (2010). In a 2006 radio interview on National Public Radio's Morning Edition, she stated, "Kids pay better attention to their subjects when they've been active. Kids are less likely to be disruptive in terms of their classroom behavior when they're active. Kids feel better about themselves, have higher self-esteem, less depression, less anxiety—all of those things can impair academic performance and attentiveness." In her interview she described a study of elementary students sharing that "groups who exercised every day performed best in the classroom and that though time was taken away from academic subjects for physical education, across the board that did not hurt the kid's performance on the academic tests. And in fact, in the condition in which the trained teachers provided the physical education, the children actually did better on language, reading, and the basic battery of tests" (Neighmond, 2006). An exciting new study by researchers at Stanford University has demonstrated the positive impact of walking on creative output. In the study published in the *Journal of Experimental Psychology,* researchers used the Guilford's Alternative Uses test to measure creative thinking. Students who took the test after walking on a treadmill and outside on campus improved their average by 60 percent—and it remained high for an extended time after sitting down post-exercise (Oppezzo & Schwartz, 2014, cited by Bartlett, 2014). In the fourth experiment on creative analogy generation, walking outside produced the most novel and highest quality analogies (Oppezzo & Schwartz).

John Dewey stated 120 years ago in his *My Pedagogic Creed*:

> I believe that the active side precedes the passive in the development of the child nature; that expression comes before conscious impression; that the muscular development precedes the sensory; that movements come before conscious sensations; I believe that consciousness is essentially motor or impulsive; that conscious states tend to project themselves in action.
> I believe that the neglect of this principle is the cause of a large part of the waste of time and strength in school work. The child is thrown into a passive, receptive, or absorbing attitude. The conditions are such that he is not permitted to follow the law of his nature; the result is friction and waste (McDermott, 1981, p. 450-451).

John Ratey (2008), an associate clinical professor of psychiatry at Harvard and Eric Hagerman, the authors of *Spark: The Revolutionary New Science of Exercise and the Brain*, shared research that emphasized how exercise enhances learning, lowers stress and anxiety, and increases cognitive performance. Their book includes chapters on the beneficial effects of exercise on stress, anxiety, depression, attention deficit, addiction, hormonal changes, and aging. In children, aerobic fitness is correlated with math and reading achievement, IQ, verbal ability, mathematical ability, academic readiness, self-control, and memory tasks.

This evidence makes a strong argument for physical education and recess in schools and for fitness programs in the workplace. In an era of cutbacks in schools of what are regarded by some as "specials" or "extras" like physical education, we are not only doing a disservice to our students' physical health and social skills development (see Harnessing the Power of Play to Teach, page 32), but we are missing an important opportunity to increase and maintain cognitive function.

This information on the importance of movement aligned with learning has implications for day-to-day practices in the classroom and workplace. Consider seating arrangements (see Designing and Re-Imagining the Classroom Space, page 57), planning breaks in direct instruction for movement, integrating movement into lessons, as well as supporting physical education in schools and fitness in the workplace. Balancing sedentary lecture time with active engagement and movement activities improves learning.

Get Them Talking and Keep Them Engaged

> The goal is the same, no matter what the age of the students; someone must be there to listen, respond, and add a dab of glue to the important words that burst forth.
>
> — **Vivian Paley, teacher and author**

We are social beings, motivated and engaged by social interaction. Positive social contact as part of learning can induce the release of dopamine and bring pleasure to learning (Willis, 2011b, 2013). People of all ages learn from coaching and influence from peers (see chapter three on collaborative learning). When learners have the opportunity to reflect and talk with peers about their lessons and experiences, they are practicing and reinforcing these lessons while making new connections as they verbally discuss their learning and hear others' insights. Social learning helps participants to synthesize the material and gain new perspectives through exploration with others (Willis, 2010a). This book explores the benefits of collaborative learning and the use of social interaction for review and reinforcement of academic content, for practicing social emotional skills, for promoting meaningful reflection, and for enhancing learning outcomes.

Play and Executive Function

To be playful and serious at the same time is possible, and it defines the ideal mental condition. —**John Dewey, 1910**

We learn some of our most basic life skills through play. Play contributes to the development of one of our most important brain functions—the ability to control and modify our behavior in order to reach a goal. Adults, adolescents, and children all learn and can potentially "change their brains" through play (Aamodt &Wang, 2011).

Research from educational neuroscience suggests that interactive social play improves executive function. "Executive function" is the term used by neuroscientists to describe cognitive processes that regulate, manage, and control other cognitive functions. These include attention, problem solving, working memory, reasoning, inhibition, evaluating, and controlling one's actions. These skills promote success in school, personal life, and work. They are emphasized in social-emotional learning and 21st century skills development programs. Poor executive function is associated with high dropout rates, drug use, and crime. Good executive function is a better predictor of success in school and later in professional life than a child's IQ (Diamond, 2010).

Developing and Reinforcing Positive Behaviors: People learn self-regulation when they practice taking turns and focusing on the different tasks necessary to participate in a game. Playing safely, staying within boundaries, following rules, and other regulatory skills are necessary for, and practiced through, play. Structured and unstructured interactive play is a powerful way to improve executive function in children and adults because many of the cognitive processes involved have plasticity and can be developed through practice. Use of these skills or mental capacities builds that ability.

Playful learning in a safe supportive environment promotes inclusion and inherently involves opportunities for adaptation and creativity. A playful approach facilitates learning by arousing attention and increasing sensory gain (Aamodt & Wang, 2011; Willis, 2010a).

Play and Dopamine: Dopamine is a neurotransmitter involved in awareness, attention, and mood. Dopamine levels are associated with the reward center of the brain and the heightened sense of pleasure that characterizes rewarding experiences. Engaging participants in learning activities that correlate with increased dopamine release will likely get them to respond not only with pleasure, but also with increased focus, memory, and motivation (Willis, 2010b, 2012).

See chapter three for more information on play and learning. Strategies to harness the power of play to teach, review, and reinforce important academic, social-emotional, and life lessons are offered throughout this book.

Reflection Creates Meaningful Connections and Lasting Lessons

John Dewey (1916, 1925, 1934) emphasized the importance of engaging learners in reflection in order to help them make connections between the learning experience, real life, and future learning. He believed that when reflective practice is part of learning, our experiences shape us, creating meaning and relevancy, and initiating further growth and change. Reflection has become a key tenet of experiential education philosophy. The modern day field of educational neuroscience reinforces Dewey's view by showing us that we are indeed shaped by our experiences and continue to be throughout our lives. Educational neuroscientists emphasize that intentional engagement in reflective practice is key to cementing learning. It does so by creating multiple pathways to learning, facilitating patterning, meaning making, and providing a forum for feedback. Reflective practice also enhances application, transfer, retention, and recall (Willis, 2013).

Brain researchers are finding that knowledge increases through pattern recognition and by matching new information to memories. This creates more extensive neural networks. We tend to remember things if we think we have seen them before. When new information is recognized as related to prior knowledge, learning extends beyond the domain in which it occurred—transferring to new learning and problem-solving situations. Positron Emission Tomography (PET) scans show that when people are given new information their brains activate the stored memory banks. When these connections are made, it becomes long-term memory (Willis, 2010a, 2013).

At the time of this writing, a team of researchers from Harvard Business School, HEC Paris, and the University of North Carolina released the results of their ground-breaking study on the

I cannot teach anybody anything. I can only make them think. — **Socrates**

positive impact of reflection on learning. Their findings suggested that reflection is a powerful mechanism by which experience is translated into learning (Di Stefano, Gino, Pisano, & Staats, 2014). They found that individuals perform significantly better on subsequent tasks when given intentional time to reflect on what they learned from the completed task. In their article *Learning by Thinking: How Reflection Aids Performance*, the researchers stated "learning generated by reflection coupled with experience will lead to improvement in problem-solving capacity as compared to learning by experience alone" (p. 5).

Meaning creates an important structure around detail. When we have association or context—an emotional connection and sense of relevancy in regards to learning—we are more likely to remember the material. The process of reflection and metacognition (thinking about thinking) enhances this process of meaning making by intentionally helping learners explore the patterns that connect the experience with real life application, context, personal meaning, and relevancy. Reflection and thinking about thinking create stronger and more numerous neuronal connections or pathways to learning. This process also provides a forum for the ongoing feedback and check-ins emphasized by educational neuroscientists as key to learning (Willis, 2010a, 2011a, 2013). Chapter nine of this book is devoted to exploring innovative reflection tools and techniques that will facilitate meaning making, ongoing feedback, patterning, and reflection.

Summary

This information about the brain and learning just scratches the surface of the emerging data from the field of educational neuroscience. Much from this field can enhance our practice as educators. Table 1.1 on the following pages illustrates specifically how current brain research connects with the tenets of experiential education philosophy and practical experiential approaches. The activities and strategies offered in this book are clearly evidence-informed, they blend neuroscience research with common-sense teaching approaches. The Activity List for Brain-Based Approaches to Learning found on pages 18 and 19 connects specific activities in this book with the brain-based approaches found in this chapter.

Table 1.1 Principles of Experiential Education and Brain-Based Learning Tenets

Experiential Education Principles*	Educational Neuroscience Research
Experiential education is active, involving learners emotionally, socially, intellectually and physically as participants in learning experiences rather than just receivers of information.	• When learners actively explore a subject, talk about it, practice it, and create meaning from it, stronger and more numerous neural "pathways" to information are formed for storing and recalling this information. • When multiple senses are used, more regions of the brain that store information are activated and the more interconnections are made, allowing content to be something we learn rather than memorize). • We are social and emotional beings. Processing centers in the brain that decide what we pay attention to and store for long-term memory are impacted by our emotional response. • Studies show that people who are active outperform those who are less active in problem solving, long-term memory, and reasoning. When physical movement is incorporated into learning situations, attention and retention increase. (Medina, 2008; Aamodt & Wang, 2011; Willis, 2010a)
An atmosphere of fun and novelty opens doors to learning. People learn best when engaged in solving novel, challenging problems that inspire creativity, questions, experimenting, and collaboration.	• Novelty, challenge, and playful collaborative approaches stimulate attention and increase engagement. The brain's sensory intake filter—the reticular activating system (RAS) through which all sensory input must pass—is particularly receptive to novelty and change associated with pleasure and to sensory input about things that arouse curiosity. • Actively engaging multiple senses (i.e., touching, seeing, exploring, talking about experiences, writing and drawing, using metaphor) stimulates different memory systems in separate brain areas, increasing long term memory. (Willis, 2006, 2012)
Relevancy and meaning are key to learning. Learners need to understand the purpose behind the lessons and feel that the material has intrinsic value and relates to real-life, present and future interests.	• In order for information to make it through the limbic "filtering" system and form synaptic connections, it needs to be relevant, engaging, and important to success or fun! Novel, relevant stimuli draw attention. From the time we are babies, we pay attention to information that is useful for survival and disregard most of the rest. • When a learner finds something intrinsically rewarding, they are more likely to engage in it and practice it. This enhances emotional connection to material and long-term memory. (Willis, 2010a, 2011a, 2013; Aamodt & Wang, 2011)
Choice and control is imperative. Learners need to feel they have the ability to make choices about their experience, take responsibility for themselves, and have a say in the direction of their learning.	• When stress is high and learners don't feel they have choices and control in a situation, the amygdala diverts information to the limbic system where primal reactions such as fight or flight or disengagement take over, blocking useful memory storage and meaningful connection making. • Neurotransmitters associated with pleasurable learning situations contribute to an increase in dopamine and norepinephrine. (Willis, 2010a; Aamodt & Wang, 2011)

Table 1.1 Principles of Experiential Education and Brain-Based Learning Tenets (continued)

Experiential Education Principles*	Educational Neuroscience Research
Learners must reflect on experience in order to learn. Reflection involves thoughtful time connecting the experience to other lessons and real life situations. Insight is the result of reflective thinking.	• Brains search for meaning to successfully encode new information into memory. Reflection promotes what neuroscientists call relational memory and patterning in the hippocampus. • When learners see relationships between previous experiences, neural connections increase and long-term memory storage and retrieval are enhanced. • Personal meaning makes the material more memorable. (Willis, 2013)
Learners thrive when they are in an environment where they feel fully valued, respected, supported and safe.	• Information can be "hung up" by the reticular activating system or RAS when participants don't feel safe and supported (Willis, 2010a). • Automatic (involuntary) fear responses take over when a person is asked to perform before they are ready, resulting in a "fight or flight" response. In these situations, learners often check out (or, in some cases, act out) focusing on their feeling of panic or stress instead of the lesson (Willis, 2013). • Pleasurable experiences, including play and positive social/ emotional connection, can increase dopamine response increasing focus, memory and motivation (Willis, 2010a). • Challenge and reward in the right amounts increase engagement (Willis, 2010b, 2011a).
Experiences should be carefully chosen to meet the differing needs and personalities of learners. This requires creativity, flexibility, and intention from the teacher/facilitator.	• Just as there are variations in the rest of our physical selves, our brains are physically different (Medina, 2008). Educators can effectively plan lessons using various methods of instruction, reflection and practice by getting to know and attending to the progress of individuals within the group. • Differentiation doesn't mean an individual lesson for each person—it means taking individual differences into account (Tomlinson, 2013).
Sequence or scaffold lessons so they can be built upon. As John Dewey (1916) stated, "Growth through experience must create conditions for future learning."	• "Learning promotes learning." People learn best through incremental achievable challenges based on the learner's ability—with ongoing feedback from their teacher and their peers (Willis, 2010). • When people practice and build their competence step-by-step with ongoing feedback they become more confident about the learning process, more comfortable with their learning environment, and more able to take on new challenges (Willis, 2010b).
More is learned through exploring and struggling than by being provided the answers. Balance spontaneous learning with guidance and role modeling. The educator initiates learning by structuring appropriate experiences—the student takes it from there.	• We learn through practice, reflecting on experiences and practicing again. Dendrites change in size and number throughout life in response to experience and practice! This ability to change through the growth and support of the connecting cells (dendrites) between neurons is called neuroplasticity. • When we feel intrinsically rewarded by an experience through mastery, feedback, and a sense of competency, dopamine is released. (Willis, 2013)

* This list of principles is my effort to summarize the tenets of experiential education philosophy based on my studies of John Dewey, Alfred North Whitehead, and Lev Vygotsy as well as my experience in the field.

Activity List for Brain-Based Approaches to Learning 

Activity/page #	Create Multiple Pathways to Learning	Emotions are Integral to Learning	Joyful Learning	Get Their Attention and Keep It	Get Them Moving	Get Them Talking and Keep Them Engaged	Play and Executive Function	Reflection/Connections/ Lasting Lessons
US List, 50		●				●	●	●
Circle 'Em Up, 63			●		●		●	
Which One?, 65		●				●	●	
Image/Object Activities (ch 5), 71-78	●	●	●	●	●	●	●	●
Journaling/Writing Prompt Activities (ch 5), 79-81	●	●		●		●		●
Group Sculpture, 82	●	●	●	●	●	●	●	●
Music Connection, 82	●	●	●	●		●		●
Graffiti Wall, 83	●	●	●	●	●	●	●	●
Concentric Circles, 86, 205	●			●	●	●		●
Handshake Mingle, 89, 143, 204	●	●	●	●	●	●	●	●
Trade and Share, 91, 146, 170, 205	●	●	●	●	●	●	●	●
Dominos and Other Simple Prop Match-Up, 95, 205			●	●	●	●	●	●
Commonalities Mingle, 97, 206	●	●	●	●	●	●	●	●
Anyone Who, 98, 205	●	●	●	●	●	●	●	●
Classroom Four Square, 102	●		●	●	●	●	●	
Pass the Knot, 103	●		●	●	●	●		●
Review and Reflection Dice, 104, 208	●		●	●		●		●
Gallery Walk, 104	●			●	●	●		●
Play Dough Pictionary, 110, 207	●		●	●	●	●	●	●
Charades Race, 112, 206	●		●	●	●	●	●	●
Bag of Nouns, 114	●		●	●	●	●	●	●
Zoom, 115, 156	●			●	●	●	●	●
Layers of the Atmosphere Line Up, 117	●			●	●	●	●	●
Dominos Fraction Line Up, 118	●		●	●	●	●	●	
Telegraph and Probability, 120	●		●	●			●	●
River Crossing Math, 121	●		●	●		●		●
Create a Colony, 123	●	●	●			●		●
Colleen's Character and Metaphor Lesson, 124	●	●		●				●
Food Web Tag, 126	●		●	●	●		●	
Active Coordinate Plane, 127	●		●	●	●			
Collaborative List Making Review, 128			●	●		●	●	●

Activity List for Brain-Based Approaches to Learning (continued)

Activity/page #	Create Multiple Pathways to Learning	Emotions are Integral to Learning	Joyful Learning	Get Their Attention and Keep It	Get Them Moving	Get Them Talking and Keep Them Engaged	Play and Executive Function	Reflection/Connections/ Lasting Lessons
Team Tally, 135		●	●	●		●		●
Have You Ever?, 140	●	●	●		●		●	
Name Activities, 143-147	●	●	●		●	●	●	
Tin Can Pass, 148			●		●		●	
Rock-Paper-Scissors Tag, 149			●		●		●	
Community Circles, 151			●		●		●	
Telegraph, 152						●	●	●
"Follow Me", 154					●		●	
Communication Breakdown, 155			●			●	●	●
Back Art, 157			●					
Back-to-Back Drawing, 158						●	●	
Fill the Crate Challenge, 158			●		●	●	●	
Turnstile, 159			●		●		●	
Turning Over a New Leaf, 160					●		●	
Newspaper Bridges, 161						●	●	
Helium Hoop, 162				●	●	●	●	●
Saving Springfield, 163			●		●		●	
Celebration/Appreciation Activities, 166-168		●		●		●		●
Personal Strengths Activities, 169-170	●	●				●		●
Group Norms Activities, 171-176		●				●		●
Time Management/Goals Activities, 177-183	●	●		●		●	●	●
Large Group/After-School Activities, 184-187			●		●		●	
Transition/Fun Filler Activities, 188-192			●	●	●		●	
Reflection Activities (ch 9), 201-214	●	●	●	●	●	●	●	●
Self-Reflection Activities, 215-216	●	●		●				●
Postcard to Your Future Self, 217	●	●		●				●

Innovative Approaches to Teaching

A Framework for Participant-Centered Learning

Snapshot: At a glance, it looks like three boys have taken over this 9th grade Earth Science class and instigated a game of Play Dough Pictionary. A closer look finds a pleased and observant teacher standing on the sidelines. The game is focused on reviewing the rock formations the class has been studying. Everyone is actively engaged in the game, students are asking each other questions about the rock formations, and the boys (who volunteered to lead) are demonstrating their knowledge of the content. A conflict about who won the last round is quickly mediated and appropriately resolved by the students. Meanwhile, the teacher is gaining insight into what the class knows about the rock formations and what she needs to review or clarify with them.

Experiential Education Connects the Dots Between

Educators often receive conflicting messages about what works and what should be prioritized in education. On one hand there are shifts toward using scripted curriculum and an emphasis on standardized test scores for accountability. On the other hand it is clear that modern day society requires an integration of a broad range of social-emotional and career readiness skills and innovative teaching approaches to prepare learners to succeed in school, career, and life. Chapter one explored the brain research that explains and supports experiential approaches to teaching. This chapter takes these connections further and makes a clear argument about why student-centered teaching that is responsive to social, emotional, and developmental needs should be a part of rigorous academic instruction or training—not separate from it. It also provides a framework to focus our efforts.

This chapter outlines an effective, holistic approach to teaching and learning by connecting the dots between experiential, brain-based education and the useful educational models of Differentiated Instruction, Social-Emotional Learning, 21st Century Skills and Strengths-Based Teaching. These pedagogical theories, validated by the brain-based principles outlined in chapter one, go beyond what some see as educational buzz words to meaningful theories supporting commonsense practice. These innovative approaches are interwoven with experiential education philosophy and brain-based research to provide a meaningful framework for the techniques and activities offered throughout this book.

Differentiated Instruction: Don't Dismiss It

If the only tool you have is a hammer, you tend to see every problem as a nail. —**Abraham Maslow, founder of humanistic psychology**

John Dewey and Jean Piaget were both early proponents of what we now call differentiation or differentiated instruction. They believed in a learner-centered approach and thought it was important to consider the unique differences of each learner who comes to the classroom—taking into account his or her own personality, strengths, abilities, challenges, and past experiences—recognizing that, because they are individuals, they will experience lessons differently. Dewey (1897, 1900) and Piaget (1926) believed good educators take this into account and design lessons in a way that allows for, and even embraces, these differences. In his theory of experience, Dewey emphasized that an experience that might be beneficial for one student could be detrimental to another. He advocated that the role of the educator was to be aware of this individuality and use this information to guide learners through lessons in a way that was appropriate for them and in a way that allowed them to take ownership and responsibility for their learning.

Differentiation has, unfortunately, received bad press or been dismissed by some in the education field because it has been misinterpreted as requiring the teacher to create individual lesson plans for each student day in and day out. Carol Ann Tomlinson (2013), a leading modern-day proponent of differentiation, defines it as a commonsense, student-centered, empathetic approach in which the teacher believes, if they know who their students are, they can see the world through their point of view. This allows the teacher to provide leadership and clarity around the goals of lessons. In differentiation, teachers believe that all students can be successful as long as they are willing to work and as long as the teacher is willing to work. Tomlinson points out the connection with brain-based research, emphasizing the importance of a safe and supportive environment along with positive emotional associations to learning. She stresses the importance of having these components in place before focusing on content. Like Judy Willis (2010b), she supports the importance of ongoing feedback and check-ins around learning progress for both students and teachers—what many call ongoing formative assessment. Tomlinson emphasizes the importance of both the educator and student understanding shared goals for learning.

Clearly the principles of differentiation overlap with experiential education philosophy and brain-based learning. All three promote the ideas that learners need to feel connected to and supported by their peers and teachers, that choice and a sense of ownership in learning is essential, and that reflection is key to learning.

The Basic Practice of Differentiation

- Know the individuals in front of you.
- Believe in your students and form caring relationships with them.
- Clarify and create goals with student input—help learners understand what is needed for a successful outcome.
- Plan curriculum that connects with something relevant in the learner's world.
- Engage actively (physically, emotionally, socially, and intellectually).
- Integrate reflection throughout.
- Regularly check in and assess learning, adjusting your plan as needed.
- Teach up with a lesson plan for the most advanced and then adapt lessons to help everyone get there.
- Scaffold lessons (see page 52).
- Vary your route, use multiple modalities to reach learners.
- See yourself as a guide or collaborator in learning and your group as a learning team.

One of the strongest arguments for differentiated instruction is based on what the neuroscience field is telling us about multiple pathways to learning (see page 7)—using different modalities to impart lessons, review material, and check for understanding. This book provides numerous activities and strategies to help educators and group facilitators differentiate or universalize learning experiences in order to meet the diverse needs of learners.

Learning for Real Life: 21st Century Skills Development

The illiterate of the 21st century will not be those who can't read or write, but those who cannot learn, unlearn, and relearn. **–Alvin Toffler, author, social scientist, futurist**

The term "21st century skills" is interpreted in different ways in the education field. Generally it is used to refer to important academic, life, and career readiness skills such as critical thinking, problem-solving, collaboration, communication, technological literacy, and understanding of other cultures. These ideas aren't newly important to society; they certainly would have been valued in many past centuries. However, in the 21st century, we do live in a rapidly changing global society and the traditional, one-size-fits-all approach to education focused on memorization won't fit the needs of a society that calls for adaptation, global communication, and the ability to synthesize rather than memorize information. Some describe 21st century skills as the ability to know how to learn and adapt to the changing needs of society. This is in keeping with John Dewey's writings:

> With the advent of democracy and modern industrial conditions, it is impossible to foretell definitely just what civilization will be twenty years from now. Hence it is impossible to prepare the child for any precise set of conditions. To prepare him for the future life means to give him command of himself; it means to train him that he will have the full and ready use of all his capacities; that his eye and ear and hand may be tools ready to command, that his judgment may be capable of grasping the conditions under which it has to work, and the executive forces be trained to act economically and efficiently (McDermott, 1981, p. 445).

As John Dewey proposed a century ago, this movement encourages educators to move away from the traditional methods of teaching that were designed for a different economic and social reality and bring into the classroom new and innovative approaches to teach curricular content along with life and career skills.

The 21st century learning movement has gained the attention of well-respected, international educators, along with business, social, and government leaders who aim to redefine how our educational systems work to meet the needs of our rapidly changing society. Bernie Trilling and Charles Fadel, authors of *21st Century Skills: Learning for Life in Our Times,* state: "To be productive contributors to society in our 21st century, you need to be able to quickly learn the core content of a field of knowledge while also mastering a broad portfolio of essentials in learning, innovation, technology, and career skills needed for work and life" (2009, p. 16).

Skills for Life

Tony Wagner, author, professor, and Innovation Education Fellow at the Technology & Entrepreneurship Center at Harvard, promotes the idea that in the 21st century we have new educational challenges that need to be addressed in new ways. He believes all students need a convergence of skills for work, continuous learning, and citizenship in what he calls a "knowledge society." In his 2011 Keynote at the Learning & the Brain Conference, Wagner stated that the current generation of youth and young adults are differently motivated to learn than prior generations, and boredom is the leading cause of low achievement and student dropouts. In the modern world the "innovators" are going to be most successful and have the most job security in the workplace. He described "Seven Survival Skills for Careers, College and Citizenship" as follows:

1) Critical thinking and problem-solving
2) Collaboration and leading by influence
3) Agility and adaptability
4) Initiative and entrepreneurialism
5) Effective oral and written communication
6) Accessing and analyzing information
7) Curiosity and imagination

Motivation

Wagner (2011) emphasizes that modern-day learners are motivated by using digital technology for extending friendships; for interest-driven, self-directed learning; and as a tool for self-expression. He believes that students learn from peers and from adults who don't "talk down to them," and that learners want to make a difference and do interesting/worthwhile work. Wagner encourages schools and educators to shift their approach:

- Create opportunities for meaningful collaboration rather than focus on individual achievement.
- Focus on creating instead of consuming.
- Emphasize multi-disciplinary learning over specialization.
- Cultivate intrinsic motivation instead of extrinsic rewards or punishments by creating learning situations that include play, passion, and purpose.

Charles Fadel (2012), in a recent presentation at the Learning and the Brain conference in Boston, made similar arguments: He proposes the rapid pace of change in our world through technology has changed the way we need to teach. The workforce in more developed countries will be involved in research, design, marketing and sales, and global supply chain management. He points out that currently in the American educational system, learning is abstract, theoretical, and organized by disciplines, whereas the modern real world workplace is concrete, interdisciplinary, and organized by problems and projects.

Innovative Teaching

Like John Dewey, Fadel and his coauthor Trilling emphasize the importance of relevancy to the learner. Their book, *21st Century Skills: Learning for Life in Our Times* (2009), argues that educators need to provide the following:

- Personalized learning
- Speedy access to online research, writing, sharing, and project tools
- Social networking and collaborating
- Playfulness, games, simulations, and creative expression

- Multiple modes of instruction
- Context-connection to the real world using relevant meaningful work, reflection, and application to real life
- Community—the opportunity to learn socially in groups and teams.

Again, like Dewey, these modern-day proponents of innovative education believe we need to create a culture and climate in our schools of high expectations, responsibility, ownership, and self-direction. A personal culture of caring, respect, trust, cooperation, and community is essential. They believe 21st century education needs to include rich, meaningful, relevant content and innovative teaching. Trilling and Fadel (2009) refer to modern-day pedagogical research in identifying five key findings from research on the science of learning. These look a lot like what John Dewey espoused:

1) Authentic Learning: Learners need to experience relevant learning with context so they can transfer lessons to real life.
2) Mental Model Building: They, like Dewey, emphasize the importance of reflection on our mental models and thinking about thinking—metacognition.
3) Internal Motivation: Like Dewey, they emphasize the importance of people needing to be intrinsically motivated to learn, and that learners need an emotional connection to relevant, interesting, and meaningful material.
4) Varying Needs of Learners: Trilling and Fadel emphasize that not all learners are the same; that we need to differentiate instruction and personalize learning to meet the individual needs of learners (differentiation, Universal Design for Learning).
5) Social Learning: They agree with this book's premise that people learn best through social interactions and collaboration.

In a nutshell, Trilling and Fadel (2009) believe like John Dewey, that the way we approach education should change with the times and evolution of society. These necessary adaptations to the needs of modern society include promoting and cultivating flexibility, adaptability, initiative, self-direction, communication, social and cross-cultural interaction, productivity, accountability, leadership, and responsibility. "The ability to work effectively and creatively with team members and classmates regardless of differences in culture and style is an essential 21st century life skill" (p. 80). For more information on 21st century learning see resources.

The Partnership for 21st Century Skills and other leaders in the 21st Century Learning movement emphasize the four C's as the learning skills that matter most:

- Critical Thinking and Problem Solving
- Creativity and Innovation
- Communication
- Collaboration

I personally have added citizenship and character as 5th and 6th "c's" and believe that the skills outlined in the 21st century skills development movement clearly connect with social-emotional learning.

Intelligence plus character—that is the goal of true education. –**Martin Luther King**

Social-Emotional Learning

Real education should consist of drawing the goodness and the best out of our own students. What better books can there be than the book of humanity? –**Cesar Estrada Chavez**

Social-Emotional Learning (SEL) is a term that emerged in the 1990s to describe the process of learning important life skills, interpersonal skills, and pro-social behaviors such as communication, problem-solving, emotional regulation, positive self-expression, goal setting, and responsibility for self and others. To me it is closely intertwined with the 21st century skills described above. The State of Illinois Department of Education, one of the first school districts to set standards for social-emotional learning, defines it as the process through which learners develop awareness and management of their emotions, set and achieve important personal and academic goals, use social-awareness and interpersonal skills to establish and maintain positive relationships, and demonstrate decision making and responsible behaviors to achieve school and life success. There is a strong research base indicating that these SEL competencies improve students' social/emotional development, readiness to learn, classroom behavior, and academic performance (Illinois State Board of Education, 2005).

For more than 30 years, numerous studies have shown that social and emotional learning interventions focused on self-awareness, self-management, social-awareness and responsible decision-making have increased academic performance, reduced aggression and emotional distress among students, increased helping behaviors in school, and improved positive attitudes toward self and others (Durlak, Weissberg, Dymnicki, Taylor, & Schellinger, 2011). The most effective interventions are those that use active-learning techniques to help students understand and practice these skills (Vega, 2014). This body of research demonstrates that a focus on social-emotional skills building is essential to a proactive, preventative approach to bullying and harassment problems in schools.

John Dewey believed that education served a broader social purpose of helping learners become productive members of a democratic society (1897, 1900). In his Pedagogic Creed, Dewey argued the value of education is not only in promoting the acquisition of content knowledge, but also providing opportunities for students to learn how to live and thrive in society. He believed the purpose of education is to help learners reach their full potential, so they could contribute to society in a meaningful way:

> I believe that the only true education comes through the stimulation of the child's powers by the demands of the social situations in which he finds himself. Through these demands he is stimulated to act as a member of a unity, to emerge from his original narrowness of action and feeling, and to conceive of himself from the standpoint of the welfare of the group to which he belongs. Through the responses which others make to his own activities he comes to know what these mean in social terms (McDermott, 1981, p. 443).

My work with local schools includes designing school action plans specifically around social-emotional learning. The SEL goals we work on in the classroom through group problem-solving challenges, insightful reflection, and group dialogue include the following:

- Decision making
- Responsibility
- Communication skills
- Cooperation
- Positive peer relationships
- Understanding social cues
- Respectful behavior toward self and others
- Interpersonal skills
- Empathy
- Conflict resolution
- Recognizing different perspectives
- Learning how to express oneself constructively
- Self-control, self-management
- Knowing how to use resources in school and community
- Goal setting
- Honesty
- Respect
- Compassion
- Personal responsibility
- Time management
- Problem solving
- Higher-order thinking
- Using different strategies
- Knowing how to ask for help and support
- Identifying and learning how to capitalize on personal strengths

These are the same life, group, interpersonal, and workplace skills that the adult training groups I work with seek to develop. Clearly they are closely related to the 21st century skills described previously. Experiential approaches give students the opportunity to practice becoming responsible members of society—helping them to learn to value themselves and others. Throughout this book, I share specific examples of experiential methods for developing these important social-emotional skills. Chapters seven and eight are devoted entirely to this subject and include a variety of games and initiatives to practice and develop these skills. More information about Social-Emotional Learning can be found in resources.

Strength-Based Teaching: The Power of a Positive Mindset

Trust people as if they were what they ought to be and you can help them to become what they are capable of being. — Johann Wolfgang von Goethe

Researchers are continuing to find that the most effective social interventions and teaching strategies are those that focus on strengths and assets over deficits. For decades, positive psychologists have recommended that educators work to build the positive attributes that help people thrive rather than focusing on the negative behaviors. Many business and education leaders have been looking in this direction in recent years (i.e., Gallup's "Strengthsfinder" [Rath & Buckingham, 2007] program and *Strengths-Based Leadership,* Daniel Pink's [2009] *Drive* on intrinsic motivation, and Carol Dweck's [2006] *Mindset: The Psychology of Success*).

Proponents of strength-based approaches have popularized the term "grit"—a positive trait based on perseverance, powerful motivation, and passion for long-term goals (Duckworth, Peterson, Matthews, & Kelly, 2007). This kind of perseverance is associated with long-term success in life. Angela Lee Duckworth and her colleagues at the University of Pennsylvania have conducted studies on the impact of self-control and grit on academic and professional success and how educators can cultivate these traits.

The strengths-based view toward youth development and education is now being recognized on the federal level. In the U.S., federal human service agencies and policy-makers are now replacing deficit focused prevention and social intervention programs that focused on negative behaviors with strengths-based approaches designed to build assets and resources in youth. The "Positive Youth Development" approach is based on longitudinal studies by organizations such as the SEARCH Institute and the National Research Council on the specific physical, cognitive, social, and emotional factors that promote positive development and help youth avoid risks and thrive to become successful, productive members of society.

This research promotes the same connecting ideas we explore throughout this book, including experiential education, social-emotional learning, brain-based learning, and differentiation. It points to factors such as the need for a safe and supportive environment, clear boundaries and expectations, a sense of purpose, competency and mastery, positive relationships with adults and peers, social skills, positive identity, and caring for and service to others.

I believe a strengths-based approach involves a state of mind on the part of the educator to believe the best about their students or participants and an attitude that everyone can succeed, they just might take a different path to getting there than planned. It is a positive approach focused on opportunities, goals, and strategies. The foundational principles of this book are based on a positive, learner-centered belief in the power of groups and individuals to engage in meaningful learning when they are supported and guided in a positive way. Chapter four explores the power of our attitude as educators. For further research around the power of a positive attitude and strengths-based approach to teaching see resources.

In summary, I contend that a focus on experiential education and brain-based learning connects the dots or ties together the best of the theories/models described in this chapter, distilling them into an emergent commonsense philosophy of experiential, brain-based teaching. This interconnected framework is the foundation and rationale for the practical strategies offered in this book. I hope this information will help you sort the variety of initiatives, theories, and models that come across your desk into a coherent and useful philosophy of effective teaching.

All of the activities in the upcoming chapters will help you teach and promote the overlapping social-emotional and 21st century skills that learners of all ages need to thrive in modern day school, work, and society. These methods inherently promote a differentiated, strength-based approach to teaching. The **Example Activities for Innovative Educational Models** list on the following page is a sampling of activities that exemplify the models described in this chapter. **Purposes/Focus** (found at the beginning of each activity) highlight the theoretical and scientific frameworks. A **Purpose List** for all of the activities can be found on page 234.

Example Activity List for Innovative Educational Models

Differentiated Instruction	21st Century Skills Development	Social-Emotional Learning	Strength-Based Teaching
Images/Postcards, 71	Toolbox, Objects or Miniature Metaphor Charms, 73	Handshake Mingle, 89	Peek-a-Who Appreciations Variation, 166
Anyone Who, 98	Trade and Share, 91	Concentric Circles, 86	Graffiti Wall, 83
Layers of the Atmosphere Line Up and Other Sequence Challenges, 117	Zoom, 115	Which One, 65	Name Meanings, 143
Dominos Fraction Line Up, 118	River Crossing, 121	Dominos and Other Simple Prop "Match Ups", 95	Appreciation Notes, 168
Create a Colony, 123	Community Circles, 151	Team Tally, 135	Postcard Strength Activity, 169
Colleen's Character and Metaphor Lesson, 124	Communication Break Down, 155	Have You Ever? 140	Trade and Share Strengths Variation, 170
Food Web Tag, 126	Saving Springfield, 163	Peek-a-Who, 145, 166, 207	Vintage Keys, 171
The Active Coordinate Plane, 127	Group Juggle Metaphor Activity, 177	Tin Can Pass, 148	Community Puzzle "Strengths Mural", 173
Collaborative List Making Review, 128	Time Management "Full Plate" Lesson, 180	Rock Paper Scissors Tag, 149	Draw Your School/Workplace, 174
Gallery Walk, 104	US list, 50	Fill the Crate Challenge, 158	Postcard to Your Future Self, 217
Play Dough Pictionary, 110	Draw Your School/Workplace, 174	Telegraph, 152	Headlines: Newspaper Article 10 Years Out, 183
Charades Race, 112	Gallery Walk, 104	Turnstile, 159	Group Norms Sculpture, 175
Bag of Nouns, 114	Fill the Crate, 158	Turning Over a New Leaf, 160	Toolbox Reflection Variation, 73, 203
Group Sculpture for Academic Review or Reflection, 82	Headlines Activity, 183	Helium Hoop, 162	
	Postcard to Your Future Self, 217	Commonalities Mingle, 97	
	Newspaper Bridges, 161		

All of the activities in this book exemplify differentiated instruction; strength-based teaching methods and ways to promote 21st century skills and social-emotional learning. This list is a sampling of specific activities from the upcoming chapters that will help you reach these goals.

chapter 3

The Power of Play and Collaborative Learning

The Why Behind Using Interactive Activities and Games to Teach

Snapshot: The desks have been pushed back from the center of the room and the students are up and moving, engaged in a group activity. They each hold a picture of items found in the various layers of the earth's atmosphere (satellites, geese, airplane, radio waves). They are working together using the pictures to line themselves up in the order of the layers of the atmosphere. There is conversation and laughter as they try to accomplish this challenge. A boy who rarely raises his hand in class has stepped up as the leader, initiating a group conversation about who goes where. As the students practice skills in a playful way, their teacher is learning how much the students really know about their earth science unit.

Playful Learning

We are never more fully alive, more completely ourselves, or more deeply engrossed in anything, than when we are at play. —**Charles E. Schaefer, pioneer in the field of play therapy**

Never underestimate the power of play in learning! Play is a serious subject. There is a great deal of literature and research on play, strict definitions of what constitutes true play, types of play, and stages of play. This book promotes playful learning practices and offers specific academic and social skill-building games and activities. My working definition of playful learning involves engaging learners in educational activities, games, and creative academic group-building projects that promote learning. Playful learning experiences are:

- attractive, enjoyable, and rewarding, including moments of challenge and frustration balanced with support, success, and joy of achievement;

- engaging and active, including movement, social interaction, and use of multiple senses;
- intrinsically motivating—inspiring control and ownership on the part of learners;
- focused on the process (and lessons learned along the way) or a means to an end versus the product or outcome of the game; and
- flexible and changing.

Playful learning situations enhance academic and social-emotional outcomes, create collaborative and supportive learning communities, instill a desire to learn, and inspire a sense of discovery. Learners of all ages can develop and practice necessary life skills through play. Well-respected proponents of 21st century learning and innovative teaching promote playfulness, games, and collaborative group challenges in teaching (Wagner, 2012; Trilling & Fadel, 2009). Play and active playful communication and problem-solving challenges are vehicles educators can use to meet the diverse needs of learners, they are tools to teach, review, and reinforce academic content in novel and engaging ways.

Harnessing the Power of Play to Teach

In chapter one, we explored how play contributes to the development of executive functions such as attention, self-regulation, and reasoning. Because play is desirable, participants are motivated to practice rules of self-regulation in ways they might not in everyday life. Games require social rules, boundaries, and structure in order to work. Lev Vygotsky, a twentieth-century Russian psychologist and proponent of experiential learning methods and the power of play to teach, argued that, in play, a child learns to follow social rules because following social rules leads to pleasure. He stated, "by subordinating themselves to rules, children renounce what they want, since subjection to rule and renunciation of spontaneous impulsive action constitute the path to maximum pleasure in play" (1933, Translation Voprosy Psikhologii, 1966).

Many of the cognitive processes involved in executive function have plasticity and can be developed through practice (see Neuroplasticity, page 6). Play, especially play that involves problem-solving tasks, social interaction, planning, reasoning, and multiple senses, encourages mental growth and self-regulation skills (Aamodt & Wang, 2011; Bodrova & Leong, 2007; Diamond, 2010). Studies on brain structure and chemistry show that play activates norepinephrine (also known as noradrenaline), which arouses attention, increases senso-

ry gain and the impact sensory information makes on the brain. It also can improve brain plasticity in some neurons (Aamodt & Wang, 2011; Willis 2010b). The conditions of play generate the signals and processes in the brain that enhance learning by arousing attention.

As discussed in chapter one, dopamine, an important neurotransmitter involved with awareness, attention, and mood, is increased when we experience something pleasurable. Playful learning experiences can trigger this "reward center" in the brain, increasing positive associations with learning, attention, motivation, and retention. Playful approaches that use multiple senses can create multiple pathways to learning by engaging students socially, emotionally, physically, and intellectually. Educational neuroscientists emphasize that the more ways in which something is learned and practiced, the easier it is to recall and access that information (Medina, 2008; Willis, 2010).

Many are concerned that play gets in the way of time spent on academics. I would argue that play doesn't have to be separate from academic or training content. Integrating play leads to more effective teaching and learning. Studies on play and learning show that children in classrooms where teachers used play as a teaching tool for sustained periods of time scored higher in literacy skills (Bodrova & Leong, 2007). Play helps educators connect with their students and learn ways to support them. Games and activities help differentiate instruction and provide opportunities for formative assessment. When using interactive activities for review, formative assessment doubles as a learning and teaching tool, an idea Tomlinson (2013) promotes in her work on differentiation (see page 21).

Play inherently involves opportunities for inclusion, adaptation, and creativity. Playful activities can bring learners of various abilities together in the classroom. When I go into my local middle school to offer community building and active academic review, we make sure all students in that grade are present, even if they are usually pulled out for special education support during that period. This initiates positive changes in the students' social interactions with each other. Students who are often excluded have opportunities for positive social interactions with their peers. Students who are traditionally known as high academic achievers and may hold preconceived notions about "special education" students gain new perspectives on the many skills and insights these students bring to class. I also see teachers gain new perspective in this way. Frequently it is a student who struggles academically or behaviorally who offers the creative practical solution to a problem or the profound reflective insight that pulls the lesson together.

This book explores creative, cooperative, challenging play. Healthy competition that emphasizes the process over the outcome is involved in many of the activities and score-keeping is de-emphasized. Play is combined with reflective practice to create lasting lessons and carry learning forward.

Why Play is Such a Valuable Tool

- Brains need breaks from focused attention like that experienced in a lecture or direct instruction (Medina, 2008). Play with a purpose can be a vehicle for giving learners "brain breaks" while still making the most of classroom time.
- Play promotes and supports positive behaviors (Diamond, 2010) and develops communication skills. Socialization is practiced and reinforced through the problem-solving, rule-making, negotiation, and planning inherent in games.
- Play helps people develop executive functions such as self-regulation and impulse control (Diamond). Focusing on the different tasks needed to successfully play a game (i.e., take turns, follow rules, play safely, stay within boundaries) promotes these skills.
- Play brings joy to learning. Research shows that pleasurable associations with learning reduce stress, increase attention and retention, and encourage future learning (Willis, 2010b, 2013).
- Play involves movement and uses multiple senses. Movement is associated with increased attention and retention (Medina). Multiple ways of practicing and reinforcing information increase neuronal pathways leading to stronger memories and recall ability (Willis).
- Play promotes inclusion. Playful, interactive activities promote social and communication skills; they are effective tools for bringing all learners of various abilities together.
- Play symbolizes the real world. Engaging in play involves modeling and practicing common, real-life social interactions and situations (i.e., creativity, adaptation, exploration of abstract themes). Creative play helps learners understand where their interests and strengths lie.
- Play develops creativity and innovation. Interactive games require problem-solving and adaptation to changing space, materials, and players—mistakes and failure are part of the process. This is a premise of experiential learning, 21st century skills, and social-emotional learning.
- Play is motivating and creates conditions for active engagement. Play in an academic setting involves group members who might not engage in more formal lecture situations. (Time and time again, I see reluctant participants unable to resist play situations and non hand-raisers actively participate in academic discussion when it comes in the guise of a game.)
- Play helps develop theory of mind—the understanding that others think differently than we do (Aamodt & Wang, 2011). Watching others at play promotes empathy and understanding different views and perspectives. Through observing learners in play situations, teachers gain valuable insights into their individual needs and personality styles.

Play as a Teaching Tool in the College Classroom

I recently ran across an interesting article entitled *Using Play to Teach Writing* by Tom Batt (2010). He used play to help students in his college writing classes see writing as "exciting, surprising, satisfying and empowering." He shared his disappointment in the students' first drafts of a basic composition. They were flat and based purely on what the students thought he wanted to read rather than what they wanted to say. This inspired him to playfully ask them to "break a rule" in their next draft. Students explored breaking rules of grammar, usage, formation, punctuation, voice, style, and audience. He observed an increase in engagement, creativity and enjoyment of the activity. Some drafts were written in landscape view, others in Morse code, and one student wrote a draft on a spiraling mobile.

Batt noticed what he called a strength of voice that arose from their increased intrinsic motivation for engaging in this way. Realizing that experimentation was okay, the students were inspired to think of writing differently. Batt believes that creativity is the essence of play and essential to learning and skill-building. He is careful to emphasize the importance of balancing the pedagogical goals of play in the classroom so there is a purpose and justification in the play with a wide margin for humor and fun—process is emphasized over product.

Note: Leslie Rapparlie's (2011) book *Writing and Experiential Education: Practical Activities and Lesson Plans to Enrich Learning* is another great resource for integrating playful learning into the secondary and college classroom.

Free Play

In the rush to give children every advantage, our culture has unwittingly compromised play. —**Alix Spiegel, National Public Radio correspondent**

Working with learners of all ages in a variety of schools and treatment programs, I notice an increasing struggle with group communication, decision-making, and conflict resolution. On the playground during recess, elementary and even middle school students come to teachers to resolve the simplest disagreements or conflicts (i.e., who will be "it" first in a one-on-one tag game). Similarly, many high school and college students struggle with group projects and the skills of group decision-making, empathy, and collaborative problem-solving. Youth often depend on teachers and other adults to pick teams for games, decide who should go first, and to resolve basic conflict situations in group work and play. These are decisions that I remember my peers and I sorting out on our own (most of the time) during our pick up games of kick the can, kick-ball, and other games we played in our neighborhood or on the playground.

As a kid, I experienced many opportunities for unstructured play in our neighborhood; multi-age pickup games of all kinds filled our summer days and after-school hours. Arguments about who was "it" or who won inevitably arose, but we figured out ways to work it out ourselves rather than interrupt the game to run home and get an adult to decide for us. Older children

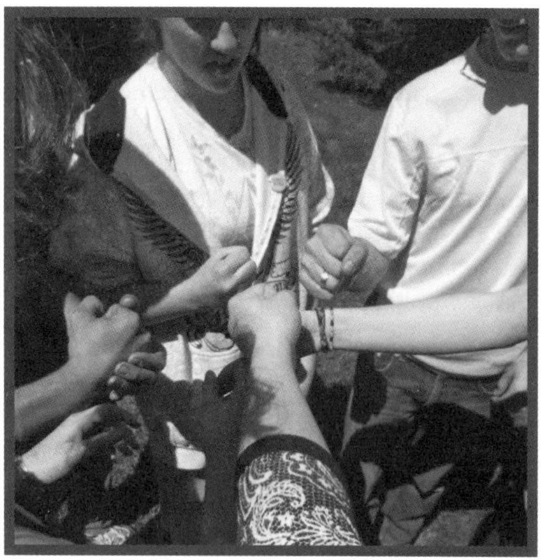

taught younger children the rules of play and acted as caretakers and role models.

Many youth in today's society don't have the same opportunity to engage in the unsupervised, free play with peers as those who grew up in the 50s, 60s, 70s, and 80s. Our society is becoming increasingly isolated. Children aren't out playing pickup games in their neighborhoods during free time after school (if they even have it) or on summer evenings. Parents are often fearful about letting kids play in the neighborhood on their own. In many schools, recess or unstructured break time has shortened or dwindled completely.

Children and adolescents participate in a great deal of solo video and computer games and online surfing, watching TV, or listening to music with their headphones. Most often children's group recreation involves team sports, clubs, and after school programs led by adults and separated by ages. Of course, computers can be useful, positive tools for learning, and there is value in structured team and after-school programs with adult role models that positively impact students. A downside of this shift, however, is that youth are not gaining the important skills that are learned from interaction with peer-only groups. They miss out on the fact that play is not "getting their way" or winning. They miss opportunities to practice flexibility and compromise in order to make a game work. In peer-only activities with little or no adult intervention, kids gain valuable experience in coming to consensus on the rules of play, decision-making, and developing social skills around communication and conflict resolution.

A recent study by psychologists at the University of Colorado demonstrates just how important free play experiences are for the development of "self-directed executive function" or the ability to set personal goals and work productively to achieve them. The authors of the study measured play habits of six-year-old children comparing structured play (adult led and organized) to spontaneous, less structured free play. They found the more time that children spent in less-structured activities, the better their self-directed executive functioning. The opposite was true of structured activities, which predicted poorer self-directed executive functioning (Barker et al., 2014).

We can help students gain these skills by providing opportunities within our adult-led, structured programs for them to practice on their own. In both my classroom programs with youth and my training work with adults, I intentionally focus on implementing participant-directed activities that practice the use of decision-making by consensus. This includes giving less direction in activities and challenges and putting more responsibility on group members for defining goals and parameters. In after school programs and during family night and field day sessions, I try to promote multi-age games sessions so older students can teach and mentor younger students. With all age groups, I purposefully create situations and opportunities that allow participants to practice coming to agreement on the rules of the game, the team name, a team symbol, etc. I use methods that involve pairs and then groups making simple choices together, practicing group consensus. This is a valuable practice/lesson for groups of all ages. As we explored in chapter one, our rapidly changing society and work world are going to require students to be problem-solvers, communicators, and collaborators.

Learning Together

As discussed in chapter one, collaborative learning is powerful. The ability to communicate and problem-solve with others is one of the most highly valued skills in today's workplace. Traditionally, our schools have emphasized a competitive approach where students try to do better than other students and/or an individualistic approach where students work on their own toward individual goals without much interaction with other students.

Sometimes we learn more from our peers than our teachers. Collaborative learning enhances positive peer impact on learning and skills development. Lev Vygotsky, an early 20th century Russian psychologist left a lasting impact on the fields of child development and education (Bodrova & Leong, 2007). He built upon the work of Piaget and other contemporaries but differed in his emphasis on the importance of the role of social interaction in learning. He believed social learning preceded development and that a learner's interaction with their community can drive learning. Vygotsky's (1933) frameworks have become a foundation of theory and research on cognitive development. His Zone of Proximal Development (ZPD) learning theory emphasized that interaction among children around tasks increased their mastery of critical concepts. This idea, introduced shortly before his death, has continued to be built upon by his followers. ZPD describes the range of tasks a child can complete independently (also referred

> In play a child is always above his average age, above his daily behavior; in play it is as though he were a head taller than himself. –Lev Vygotsky

to as the child's actual developmental level) and those they are only able to complete with the guidance and assistance of a more-skilled individual (also referred to as the child's potential developmental level).

From my own experiences as a teacher and learner, I believe this theory can be applied to all learners of all ages. I think of the many examples of this kind of learning from peers in the professional setting and try to cultivate peer teaching as much as possible in my professional development workshops and courses. This theory is seen as a tool for educators to initiate tasks that have the appropriate level of challenge to increase positive learning. "Appropriate" in this case meaning just challenging enough to stretch learners and not so challenging as to cause frustration and disengagement (see sequencing and scaffolding, page 52).

Like John Dewey's work, Vygotsky's theories (1933) fit with what we are learning from the field of modern day educational neuroscience. Vygotsky's work supports the ideas of collaborative learning and the importance of peer support. Collaborative activity among peers of any age promotes growth. Children and adolescents of similar ages are likely to be operating within one another's zones of proximal development, as college students or colleagues in a similar field would. This means that, in the collaborative group, people will model more advanced behaviors than they would perform as individuals. Collaborative learning is most effective when group goals are clearly established, reflection is an ongoing part of the process, and individual accountability within the group is emphasized as an important part of the collaborative process.

The Value of Collaborative Learning

- Social interaction enhances engagement and learning. We learn through conversation, emotional engagement, articulating and defending ideas, posing and answering questions, and learning about others' experiences and perspectives.
- Collaboration with others enhances the generation of ideas, creative problem-solving, and depth of understanding.
- Positive peer activities impact learning and skills development. According to Vygotsky's theory (1933), interaction among children around tasks increases their mastery of critical concepts—children model more advanced behaviors in the collaborative group than they perform as individuals.

- When individual accountability is emphasized in a group experience, learners practice responsibility for their actions, negotiation, conflict resolution, and understanding the impact of their behavior on the community.
- Ongoing group reflection is integral to learning from varying perspectives, evaluating and expressing feelings, and transferring the learning from the experience to future learning.
- Students feel the joy of reaching a common goal when collaborative learning involves time for celebration of successes through reflective activities or a formal "unveiling of the work" (see Snapshot on page 1).

Sources of the Games and Activities in This Book

As an experiential educator and recreational therapist, interactive activities and group challenge initiatives are the "tools of my trade." The activities in this book come from my experiences studying education, recreational therapy, and adventure education in college and graduate school, through books, from conference workshops, and through co-leading with other facilitators and educators. The best techniques have arisen from the ongoing co-creation that happens when you experiment with a variety of group situations—sometimes on the fly when you didn't have all the materials you needed. In these moments you learn new twists and adaptations from your participants and through experimentation (a lot like cooking).

The New Games Foundation was the main resource for activities during my first experiential education job in 1989. At the time I worked in a collaboration program between the YWCA and a school district's special education program using games to teach social skills in the classroom and after school programs. *New Games* and *More New Games* (Fluegelman, 1976, 1981) were my "go to" resources. Many of my tried and true favorite activities, and my attitude and orientation around play as an inclusive and empowering learning tool originate from the New Games Movement. The New Games Foundation is now defunct, and its books are out of print, but the spirit and the games live on in the modern day fields of team-building and therapeutic recreation. The original books as well as recent articles and blog posts about the movement are worth searching out (see resources).

As a recreational therapist in a clinical treatment setting, I learned a great deal from co-leading and experimenting with a group of talented expressive therapists from a wide variety of backgrounds (art therapy, music therapy, recreational therapy, and occupational therapy). They all brought creative activities and adaptations to our work with patients and family groups. During my time in adventure education, I had the good fortune of working for Karl Rohnke, one of the most prolific disseminators of interactive group problem-solving games and challenge activities. You will see many references from those sources here.

Through the years, I put my own spin on these activities, inspired by the many groups with which I have worked. In the classroom, I made specific adaptations with academics, differentiation, and social-emotional learning in mind.

Original sources for a game can be difficult to pinpoint, as the games change with each facilitator. I have done my best to cite the source where I first learned the game and to honor that person's contribution to the field. I do this to demonstrate how educators learn from each other and how games evolve. Most importantly, I aim to give you additional resources for searching out games and activities that you can use to improve your practice.

Even if the game is familiar, you can learn creative new facilitation twists, new ways to enhance buy in and inclusion, and/or innovative ways to use activities for multiple purposes. Hopefully you will find creative inspiration from the stories about the discoveries I made while facilitating in the classroom, boardroom, and therapeutic setting.

More Than the Games

Facilitators often look to add games and activities to their toolbox. I hope you will find some here, but I also hope this information will encourage you to think more about the WHY behind the activities you are facilitating rather than just adding to your repertoire. The value in using activities to teach lies in understanding the situation and process of the group's engagement and reflection. The activity or group challenge is just a vehicle to build community, practice skills, build rapport, get people moving, and bring joy into learning.

Educators can get caught up in wanting to play their favorites games without reflection—placing the focus on having the game look right (i.e., following the rules, finishing the game), rather than on what the group needs. They miss the teachable moments found in the PROCESS rather than the outcome. Remember it is not the game, but the care in facilitation, observation of the group, and the time spent on reflection that makes the game worthwhile!

Build a strong foundation of understanding, empathy, and trust within your groups and empower learners with choice and control. This foundation combined with thoughtful sequencing of activities to enhance learning and ongoing reflective practice is the key ingredient in recipes for successful learning outcomes.

chapter 4

The Inspired Educator

The Art and Science of Teaching and Group Facilitation

Snapshot: The two student-life program leaders have spent a lot of time and effort designing a facilitation skills training for leaders on their college campus. They carefully sequenced activities to engage the group, role model quality facilitation, and add to their staff's toolkits. From the beginning of the program, they encouraged participants to share their insights and ideas, demonstrating a participant-centered approach. They blended reflection and check-ins throughout. During one of these reflective activities, it becomes clear the group wants more discussion of the challenges that arise in facilitation rather than more games to add to their toolkit. The program leaders respond by putting aside some of their agenda. They change their plan midstream, asking the group to brainstorm a list of the topics they want to explore. Then they engage the group in a Gallery Walk activity using these topics and questions (see page 104).

Tips, Tools, and Mindset

There are two kinds of teachers: the kind that fills you with so much quail shot that you can't move, and the kind that just gives you a little prod behind and you jump to the skies. —**Robert Frost**

Teaching and group facilitation are both an art and a science. Our style as educators impacts the groups we work with as much as the activities or curriculum. How we introduce activities, form groups, co-facilitate, and sequence activities are as, if not more, important than the activities we choose. This chapter explores the tips, tools, strategies, and mindset that make teaching and group facilitation work. I share the strategies that I and many of my colleagues have found useful for managing large groups, dividing teams, designing the classroom or group room space, working with other educators, cultivating creativity, and engaging group members. These practical suggestions will help you plan and sequence your lessons to create ownership and buy-in with group members.

Attitude is Everything

Hope is patience with the lamp lit. —**Tertullian, first-century Latin writer and Christian theologian**

Learners truly respond—even subconsciously—to an educator's attitudes, demeanor, and expectations. We often communicate more than we realize with our body language and tone. A positive attitude is contagious. If you really believe in the methods and activities you are using, participants will most likely buy-in and respond to your enthusiasm. Conversely, if you aren't comfortable with the material, they can sense as well. Learners also respond to your attitudes toward them.

Remember to acknowledge and believe in the abilities of the learners in your groups. At times, you may work with difficult groups and be confronted with difficult situations. Be aware of how you respond to the challenging behaviors that arise. If you find yourself focusing only on the negative behaviors, or if you are starting to feel there is "no hope for that kid" or a participant is "pushing your buttons," then it might be time for some self-reflection. Take a step back to reflect on the positive aspects and achievements that have occurred; recognize the small steps and successes of the individuals and the class as a group. It could be that it is time to get some support and new perspectives from a colleague. If possible, ask a fellow educator to co-teach or observe and share his or her insights into the dynamics of the group and some possible new solutions.

During a recent conversation with a group of teachers about the challenges of working with middle school students, I said, "When I work with that 7th grade group, I keep thinking of the comedy film *What about Bob?* (Oz, 1991), where the main character (played by Bill Murray) was guided by his psychiatrist (Richard Dreyfus) to repeat the mantra 'baby steps' as he got over his phobias. I try to recognize each 'baby step' the group members are taking. If I don't, I could become very discouraged." One of the teachers, Greg Gammons of Lincoln-Sudbury Regional High School responded, "Yes, we have to remind ourselves that we are looking for an oak tree kind of growth rather than expecting mushroom kind of growth." Greg's comments have stuck with me over the years. I have observed that educators who regularly work with challenging populations are skilled at acknowledging and celebrating the small successes and adept at noticing the small, positive steps forward students are making. Research on the brain and learning emphasizes the value of repeated practice for actually changing brain functioning over time. Practicing and re-practicing skills over time does lead to an oak tree kind of growth!

All of us occasionally question our effectiveness. Teaching and group work is not easy. It takes a great deal of energy and commitment, as well as a willingness to take on challenging situations. Teachers who continue to enjoy their work for the long term remain hopeful, keep their perspective of the big picture, and recognize that growth and change often arise from the more difficult struggles with lessons. Consistently practicing patience, empathy, and the power of positive thinking reaps great rewards (Stanchfield, 2007).

Adaptation

A pessimist sees the difficulty in every opportunity; an optimist sees the opportunity in every difficulty. —**Winston Churchill**

Adaptation and improvisation are what make play powerful. The beauty of the interactive and collaborative activities in this book is that they are ever changing with each group. You will notice that in the activity descriptors I use the term "facilitation suggestions" instead of "directions" because there is no one set way to play. The game is just a tool to build community, ignite learning and fun, spark reflection and dialogue around a subject, or practice and reinforce important skills.

Be open to adaptation and aware that your group members will help you adjust to their varying needs. My ongoing experiential activities course for high school advisory groups includes a person who is visually impaired and another who is hard of hearing. We experiment with different ways to make games accessible to everyone in the group, generating significant discussions on the varying abilities and differences of the individuals in our classrooms that should be taken into account as we plan activities. We have gained an increased awareness of the components of the activities we are planning and are prepared to adapt them to fit our groups. At the same time, we are learning not to be afraid of initiating the activity because of varying abilities in our group. Our participants will help us figure out ways to adapt and include everyone. The best tips, variations, and games in my toolkit have

come from situations in which the group needed something different than what I had planned. Group members asked for or initiated an adjustment to the game or activity. This is the beauty and power of play and learning from experience.

Allowing for Labor in Learning

We only think when we are confronted with problems. — John Dewey

The art of teaching and group facilitation requires a careful balance of challenge, observation, encouragement, guidance, and the ability to know when to step in to help learners and when to step back and let learners struggle through a problem. Whether I am facilitating in the K-12 classroom, a team-building situation, or organizational group, I often see educators or group leaders struggle with allowing learners to labor through difficult problems. When learners are engaged in problem-solving activities, the adults or leaders in the room often try to direct the process or solve the problems for them. When this happens, opportunities for participants to gain valuable life and academic skills on their own, or with their peers, can be missed.

Educators are trained to help learners. Sometimes they are uncomfortable when participants grapple with a problem or the group process looks messy or disordered. They may jump in to help prematurely because of perceived/real time restraints rather than the group's actual distress. This leads to missed opportunities for learners to take ownership and control over their learning. With that said, there are also times when struggling or laboring through a problem becomes so difficult that meaningful learning stops as frustration and/or anxiety grows. John Dewey (1938) referred to these as "miseducative" experiences—experiences that get in the way of growth or further education. For example, a student becomes overwhelmed or frustrated to the extent that he/she dismisses the entire lesson. Navigating this delicate balance between group members' learning from struggling through a problem and being paralyzed by frustration is where the art of teaching comes in.

Educator as Guide

The greatest sign of success for a teacher ... is to be able to say, "The children are now working as if I did not exist." — Maria Montessori

I often find myself encouraging educators to reflect on the idea of a "learner-centered" view of teaching and group facilitation. Using this approach, educators aim to gradually shift responsibility for the success of the learning experience from themselves to the learners. As learners progress, the educator fades more and more into the background, allowing learners to take more ownership and control over their learning.

This balance has to be discovered and maintained with each learning situation and group of learners. In the learner-centered approach, educators pay attention to the progress of their groups. They put forward as many questions as answers. They learn to be comfortable with the "messiness" or frustration that is necessary in some group-learning situations, and they are willing to carve out time for a meaningful, student-centered resolution to come to fruition. They carefully observe and facilitate opportunities for ongoing reflection and feedback, so they know when they need to adjust their role to fit the needs of their group.

For interesting reading on this subject of allowing for struggle and labor in learning, see Plato's Theatetus. One of Plato's dialogues focuses on the nature of knowledge, using the metaphor of the teacher as the "midwife" of ideas. This dialogue between Socrates and Theatetus, compares Socrates' role as a teacher to that of a midwife. The midwife acts as a guide and support. She cannot go through the labor for the mother-to-be; she can only help, encourage, and guide the woman through the process. She only intervenes with the labor when the situation becomes dangerous or too painful. An effective teacher similarly practices the art of the guide, coach, and supporter in the delivery of learning, but knowledge and understanding are the labor of the student. Socrates states: "The many admirable truths they bring to birth have been discovered by themselves from within. But the delivery is heaven's work and mine" (Cornford, 1934).

When learners feel responsible for their learning and have achieved a challenge through their work, they are empowered to take ownership for future learning, growth, and change. Knowledge gained through such meaningful experience becomes lasting knowledge. As John Dewey stated (1897), "The teacher is not in the school to impose certain ideas or to form certain habits in the child, but there as a member of the community to select the influences which shall affect the child and to assist him in properly responding to these influences."

Principles of Learner-Centered Teaching and Group Facilitation

- Create opportunities for learners to take leadership and responsibility for decision-making within lessons or group experiences. Participants will then gain a stronger sense of personal accomplishment for their successes.
- Initiate learning, and allow students to take it from there.
- Structure appropriate experiences and lessons, yet remain flexible, acting as guide and role model.
- Build trust by encouraging participants to share at their own pace. Allow participants to volunteer and to "pass," especially at first. Involve participants in partner and small group discussions before sharing in front of the whole group.
- Extend a sense of choice and control to the learner, creating buy-in and ownership over learning.
- Be willing to let go of your agenda to meet the needs of the group. This might seem tough in the age of standardized testing, but within the frameworks of standards or goals, weave in opportunities for flexibility. Pushing through an agenda that doesn't meet the needs of the group will surely be a waste.
- Encourage spontaneous learning; when possible, "go with the flow" and move with the lessons the group is creating.
- Be prepared for unexpected learning opportunities, and welcome teachable moments that arise when lessons go differently than planned.
- Be prepared to artfully navigate back to the present moment in order to meet the group's needs or curricular content.
- Make learners feel valued, respected, and supported.
- Direct questions back to the students or group members. Encourage participants to help each other.
- Use activities that do not involve you as the teacher or facilitator by dividing the group into smaller reflection groups. During processing/reflection discussions, the educator or group leader doesn't have to hear everything said in the discussion for it to be effective. Sometimes learners will share more in smaller groups without a leader present. Trust the learning process!

The Collaborative Educator

There are two ways of spreading the light: to be the candle or the mirror that reflects it.
—Edith Wharton

Practicing What We Preach

Throughout this book, we explore building a positive and supportive environment for learners. I would argue, especially in schools and other long-term educational programs, this process begins with the relationships between the educators that make up the teaching team. The adult staff sets the tone in a K-12 setting. In college settings, students are aware of the climate in their department. Students pick up on our examples. If we are promoting social-emotional learning and community building, then we must practice what we preach by intentionally developing positive and supportive relationships with the colleagues that make up our teaching team. The arguments supporting the rationale for students to spend time creating community/teambuilding apply to educators as well.

Teachers often comment that the time spent participating in community building with their colleagues during an in-service day or ongoing professional development program was as valuable, if not more so, than the specific tools and activities we explored in the training. In the fast-paced environment of K-12 schools, teachers can become isolated in their classrooms with little opportunities for communication and collaboration with other teachers. In the college setting, many instructors primarily work alone. Staff or departmental meeting times are often rushed and more informational than collaborative. There is great value in taking part in staff activities that build rapport, knowledge of resources within the group (e.g., networking between subject areas), and an atmosphere of respect and understanding.

It does take a commitment on the part of the administration and willingness on the part of individual teachers to reach out to each other and find ways to work with colleagues of other disciplines and grade levels. Many educators I work with share that some of the most meaning-

I often facilitate an activity called **Zoom** (page 115) during faculty team-building sessions. Participants work together to line themselves up in the order of their pictures (each person holds a piece of a sequence) without showing them to each other. Teachers do a great job of finding others in the group with pictures similar to theirs, but then they get stuck in that small grouping and seem to resist the idea of gathering the entire group together to share the big picture and solve the challenge. More than once, a teacher has said something like "This is what happens at school. We get caught up in our own little world or grade level or subject area, and we have difficulty relating to the other groups of teachers. We need to be aware of the big picture and see ourselves as a team sharing a vision of teaching an entire body of students."

ful and rewarding moments in their careers have been when they were involved in co-teaching or cross-curricular projects.

Some schools have taken a cue from the business world and formed Professional Learning Communities (PLCs) and/or action research groups. These efforts are based on the idea that professional practitioners are in the ideal place to actively study and improve their profession. Knowledge of the field lies in the day-to-day experiences of educators; and when this knowledge is shared and understood through critical reflection among peers, practice improves and learning outcomes are enhanced (Anderson, Herr, & Nihlen, 2007). A number of studies have found that PLC's positively impact teaching practice, school culture, and student achievement (Vescio, Ross, & Adams, 2008).

As models for professional development, the action research movement and professional learning communities motivate educators by giving them an active role in identifying the purpose of their work. Rather than relying solely on solutions and expert knowledge from outside they are involved in recognizing the wealth of expertise and knowledge they have to offer each other (Stanchfield, 2007; Zeichner & Liston, 2013). This insider knowledge is what Donald Schon called "knowledge in action" in his book, *The Reflective Practitioner* (1983).

Some schools make this kind of collaboration part of a regular meeting, in-service, or release time. Some make time for team building at the beginning and throughout the school year. A number of teachers I work with have carved out collaborative team-meeting time within the structures of their schedules. If this is not the case in your school, small things like making a point to share your lunchtime with a colleague you rarely see, sitting next to someone different at a staff meeting, or visiting a colleague's classroom and asking how things are going can cultivate a culture of positive collaboration. Not only will these efforts help you develop satisfaction in your own work, your students will take note! Remember, we teach by example.

Co-Teaching

In my own experience, and in talking with other educators, I find that the practice of collaborative teaching can be the most rewarding and, at the same time, challenging aspects of our jobs. Two educators attending a course I taught, shared how they joined classrooms, creating an opportunity for co-teaching. Although it was exhausting at times, it was a highlight and positive turning point in their careers. They learned more from the process of co-creation and sharing ideas during that year than they had in the previous 20 years of teaching. It validates my own experiences, and how I have learned most of my favorite activities, methods for introductions, and fun stylistic twists from other facilitators and teachers.

There is a saying, "We teach what we need to learn," attributed to Richard Bach. When we have to explain our ideas and plans to someone else, it helps us reflect on our rationale and provides insight into our practice as an educator. This reflection pushes us to look at our strengths and weaknesses while articulating our methods. Working out plans and differences in philosophy and approach, though at times challenging, has been incredibly rewarding and definitely has improved my practice. Educators who work alone often feel stuck in a rut, relying only on their own view. If you do not have opportunities to co-lead groups or co-teach in your daily practice, try to create those opportunities by co-leading a conference presentation or an in-service with another staff or faculty member.

Based on what we have explored about research on the brain and learning, cross-curricular teaching can greatly benefit learners. When cross-curricular connections are made, learners are more likely to retain and apply the information. Learners benefit from the combination of two or more teaching styles. Educators are often more effective as a team and better able to capture the teachable moments that arise in group facilitation. The varying strengths and perspectives each facilitator brings to the group can make the experience more interesting and rewarding.

It is important for co-teachers to take time to reflect together after a group experience. Journaling and keeping a record of activities helps co-leaders plan, organize, and follow up together. Reflecting afterward, in an atmosphere where one can safely celebrate successes and communicate concerns and frustrations, helps enhance collaborative work. By opening ourselves to working with others, we are creating a richer experience and, in many ways, being "truer" to the concept of interactive group work.

Tips for Positive Collaboration

- Take the time to plan and communicate. Ask questions; talk about expectations and clarify goals beforehand. This seems obvious, but we often rush through these moments and just "wing it." Conflict can result when two leaders get together to facilitate or teach and realize they do not have a shared vision.
- Get to know each other's style. Visit each other's class or group to better understand each other's perspectives.
- Be willing to let go of your way of doing things. Compromise is one of the skills we are trying to role model for learners, right? You can always learn something from a co-leader, especially if you are willing to take a step back at times.
- Take time to reflect with each other about what worked well and what you would do differently next time. Make sure to celebrate your successes. Take a look at the reflective educator questions starting on page 222 as a guideline.

Exploring Healthy Group Norms

Wherever we look upon this earth, the opportunities take shape within the problems.
—Nelson A. Rockefeller

Group norms are the values and characteristics that exist in every group, good or bad. They include code of conduct, behaviors or customs, habits, and expectations about how things will be done. Group norms influence how group members communicate and work together. Creating a formal time and space to explore existing group norms for behavior and language is part of giving the group control over its learning experience. When a group intentionally works together to agree on clarifying and establishing norms that all members can endorse, the group becomes stronger, more effective, and more aware of its dynamics and behavior (Stanchfield, 2007). In my experience, it helps to do this after group members have spent some time together and have an understanding of their dynamics and what they will be encountering together. A meaningful group norms agreement is ongoing and should be reflected upon and revisited throughout the year or program. Chapters eight and nine offer a number of activities and reflective strategies that focus on building positive group norms. An example of a simple group norms discussion prompt follows.

US List

Purpose/Focus: group norms discussion prompt, social-emotional learning, 21st century skills, community building, citizenship, self-regulation, responsibility, collaboration, communication, choice, control and ownership, empathy, gossip and rumors, goal-setting, trust, peer interaction, emotional connection to learning, mindset

Material: large sheet of paper, marker

This activity can spark discussion around positive group norms and expectations. It can highlight key issues such as the power of language and behavior in the group process. This is best done near the initial stages of group formation but after the group is starting to build some comfort with each other. It can be helpful when a conflict or negative behavior first arises (see other group norms activities beginning on page 165).

Facilitation Suggestions

- Take a large piece of paper and write "US" on the top of it.
- Ask participants to list the characteristics and behaviors they want to have in their group (i.e.,

Knowing and Using Names in a respectful way builds trust and positive communication. Activities that clarify names can be helpful even when participants think they already know each other. A person's name is important to them and should be honored with correct pronunciation and proper use. Weave name activities and practice into introductory activities. Effective methods for introducing and reinforcing names can be found starting on page 142.

What are the ideal behaviors we want to show? What do they look/sound like?).

- I used to divide the list into two columns and ask participants to list the characteristics and behaviors they want to keep out of their group under "NOT US." I have moved toward focusing on the positive instead, using only the "US" list. For example, if "no name calling" is suggested, I explore with the group what this means and what the positive opposite is, (e.g., respectful, kind, and supportive language).
- Regardless of the method used, the group norms agreement is most useful when it is placed where the group has access to it and can refer to it and refine it during the time they work together.

Outcomes/Processing Ideas

Focusing on the positive truly does lead to positive change. I have found as the group develops, the need for the NOT US part of the list falls away if there is enough focus on the US side. Letting go of the negative allows the positive to shine through.

When initially involving a group in this activity, I may use both the US and a NOT US lists and ask where sarcasm belongs. This leads to interesting conversations. Often the group doesn't come to a conclusive decision about it at this stage, but the fact they start talking and sharing about the issue seems to make them more aware of the power of their words.

This can be a great opportunity to talk about showing appreciation for group and individual success. Have members talk about how they can demonstrate positive appreciation.

Often groups just need more information about and practice with using positive language and behavior. In school and treatment settings, kids are bombarded with information about what they shouldn't do. It is nice to explore what they can do for a change.

Reference: Adapted from *Tips & Tools* by Jennifer Stanchfield. This activity was adapted from my work at Bridgeport Public Schools during my time at High 5 Adventure Learning Center.

The Right Ingredients at the Right Time

No one is born a great cook, one learns by doing. —**Julia Child,** ***My Life in France***

As with cooking, teaching is an art that involves a combination of practice, observation, knowledge of theory, and creativity. Effective group facilitators and teachers resemble a good chef, adding together the important elements in the right amounts at the right time to create a palatable and hopefully meaningful experience. Through careful observation of all the elements involved in a group's personality and setting, the facilitator can intentionally choose and order activities to maximize learning opportunities. Many educators call this important aspect of facilitation and teaching "sequencing."

Sequencing involves consciously and thoughtfully presenting activities in a specific order to maximize learning outcomes and maintain the emotional and physical safety of the group. Being observant and intentional in your planning, presentation, and evaluation of activities are essential aspects of effective education. Though sequencing is purposeful, an educator must be flexible in moving with the needs of the group and balancing the essential ingredients with adjustments based on the readiness of the participants.

Instructional **scaffolding** is a similar idea promoted in the education field to provide support for learning when new ideas or skills are being introduced to a learner. This learning theory was introduced in the late 1950s by the cognitive psychologist Jerome Bruner who based a lot of his work on Lev Vygotsky's ideas. Scaffolding is a temporary framework used to support learners taking on a new skill or task. This support could range from coaching to providing specific directions to assigning a task for practicing and developing skills. These supports are gradually removed as students start developing their own learning strategies. The art of teaching is knowing when a skill or content you are teaching needs scaffolding (support), when is the best time to implement this support, what kind of support is needed, and when it should be removed.

There is no one specific "recipe" for sequencing or scaffolding activities or lessons that fits every group. In cooking, there is a lot of room for creativity and adaptation but there are key rules and fundamental ingredients needed in order for a cake to rise or a sauce to thicken. The same is true for the facilitation of learning experiences. Approach sequencing as a dynamic process that takes into careful consideration the personality and dynamics of the group, your strengths and style as an educator, participants' emotional and physical safety, the group's goals and agenda, available activities and materials, allotted time, and physical environment.

The timing and ordering of ingredients is key to the success of many dishes. Ingredients often need to be

added in a certain amount, a specific order, and then cooked for just the right duration for best results. Effective facilitators are in tune to the importance of activity choices and the ordering and timing of experiences. They pay attention to the group development process and allow time for trust building. It is important to balance the level of the activity and/or challenge presented and the participants' ability to meet the challenge or activity. Educators need to continually observe their group in order to be sure the selected activities fit the needs and goals of the group and the specific situation, at the same time, staying aware of and sensitive to individual group member's needs for scaffolding support.

The time needed for participants to create relationships and build trust is different for every group. When interpersonal connections and a sense of community are developed, participants will take learning further and get more benefit out of the activities in which they engage. Allow time for this to happen by choosing activities that build upon each other.

When sequencing learning experiences, be sensitive to the time of day and the physical comfort and attention span of group members. Flexibility in dealing with the unexpected is key. Listen to your group and be prepared to change your plan midstream in order to adapt to the ever changing needs of the group and to take advantage of new opportunities for learning that emerge as a group works together.

Careful sequencing and scaffolding maximizes participation by allowing people to engage at a pace that works for them. Experiential group work can be very powerful. If groups are ready to engage in the process great things can occur. Conversely, if a group is not emotionally or physically ready to encounter certain learning adventures, the experience could be damaging or inhibit growth and learning.

Sequencing and Scaffolding Suggestions

- Be prepared with a continuum of activities that build upon each other. Having activities in your "back pocket" allows you to be ready to deal with changes in direction and learning opportunities that arise in an ever-changing group.
- Be flexible enough to throw out or let go of that well-developed plan if the group's needs are different than expected.
- "Indicator" activities are helpful. Know some activities that help you read and evaluate the group and introduce challenges incrementally.

- Let participants know what is expected of them and the type of activities in which they will be participating. Informed consent is critical. It doesn't have to give away the novelty of your approach. Students who choose not to participate will at least know what they might be missing.
- Continually observe your group and re-evaluate your plans in order to be sure the activities fit the needs and goals of the group and the specific situation.
- Be sensitive to the time of day and physical environment when presenting activities.
- Take time to build relationships and trust between group members.
- Be prepared for the unexpected.
- Take advantage of teachable moments. Ongoing processing or reflection is key to moving learning forward.

Recognize each group is unique and participates in activities in a different way. Even when working with groups with similar characteristics, in the same setting, with the same program goals, I have found the actual lesson plan changes with each group in response to that group's particular personality and needs. Activities you carefully plan prior to a class or group session may be specifically relevant for one group's personality and needs and not another's. This is one of the exciting aspects of group facilitation and teaching.

There is great variety in group experience and in varying opportunities to facilitate learning. With experience, educators develop the art of reading their group and adjusting activities in a creative way throughout the group process to move learning and change forward. This is the beauty of experiential education, nurturing spontaneity of experience to take advantage of teachable moments. The art is in balancing this spontaneity and creativity with the key ingredients and timing to make it all come together successfully.

Choice, Control, and a Sense of Ownership

When a leader's best work is done, the people say "we did it ourselves". —Lao-Tzu

Ownership and Buy-In

One of the tenets of experiential education is that people learn best when they perceive a sense of control and have choice and ownership in their learning experiences. Think about creating opportunities that build this sense of choice and control for participants from the very beginning of the program or school year. Empowering learners to set reasonable parameters around their participation creates an atmosphere of healthy trust and will increase involvement from reluctant participants. Create opportunities for participants to make choices within an experience. Consider some of the following techniques:

- Add rules to an icebreaker that allow the "it" person a way out or an option to participate at his or her own pace.

- Invite participants to volunteer rather than calling on them to share.
- Allow participants to pass during group discussion.
- Help learners understand the purpose of the activity and why the lesson is relevant and useful to them.
- Make time for both the educator and learners to share expectations and goals for the class or lesson and regularly check in throughout.

Involving Reluctant Participants

Creating buy-in is a key ingredient in group facilitation. Participants can be reluctant to participate for a variety of reasons ranging from past experiences; anxiety; a developmental, physical, or mental health condition; relationships with other group members; and/or other reasons unknown to the facilitator. As a facilitator you can help people engage meaningfully in a way that is just right for them:

- Find something to intrinsically motivate group members and increase buy-in. Some people will jump at the opportunity to help set up equipment or be involved in some supportive role such as group photographer.
- Focus on positive participation. Give those who are opting out the opportunity to participate passively (that does not mean distracting the group). A "low stakes" or observational role like the photographer or note taker/reporter can draw in participants challenged by more active roles or group situations. Once you draw a critical mass into the group activities, more will follow.
- Have flexible expectations. Acknowledge the need for taking "baby steps." Group work is a process, not an event. Learning to play and be part of a group often takes practice. In working with youth, I have discovered that some actually never learned how to play and many students are not familiar with working in groups.
- Recognize that people learn and are more comfortable interacting in different ways. It is essential to differentiate the way you present material and engage group learners. Take brain research into account and design lessons that use multiple senses and involve movement and social interaction. Activities using different senses and requiring different skill sets will reach more students.
- Keep it interesting. Props, humor, and relating activities to popular culture are useful strategies for increasing involvement.
- Use peers as role models and leaders where appropriate.

Many times participants who are initially hesitant, resistant, or "too cool for school" at the beginning of an activity or program eventually become the star of the show. Over and over, I've seen learners who did not engage in class or perform well in previous group situations excel in experientially-based group work. This was especially evident when they felt empowered by having choices about their participation and were motivated by intriguing challenges. Giving them this control and choice empowered them to eventually join in despite themselves! It appeared harder to sit and watch peers enjoy an engaging activity than to join in.

Some of these students were attracted by the offer of helping me with equipment or by the possibility of being "judge" during a game. Students who were initially the most reluctant and resistant volunteered to facilitate in extracurricular programs with younger students. I couldn't count the times I have heard teachers/group leaders say something like, "I can't believe how well he participated in the activity today; he is never like that in class" or "She never talks in class; I couldn't believe she actually led that activity!" The lasting lesson for facilitators is to find ways to help participants and group leaders transfer these successes back to day-to-day participation in school or other life activities (Stanchfield, 2007).

The Practice of Ongoing Reflection

The goal of education is to enable individuals to continue their education. —**John Dewey**

Throughout this book there is an emphasis on the importance of reflection in learning. Reflection can become an engaging part of learning and teaching instead of what is sometimes perceived by learners (and facilitators) as a boring follow up to an engaging experience. I encourage educators to think about the language they use around reflection, the timing of reflection, and their attitudes toward reflection. Traditionally, many educators think of reflection as a follow-up to an experience and use the term debrief to describe this post experience wrap-up. Educators should instead view reflection as an ongoing, engaging practice that begins the moment learners walk in the door (or maybe even before with pre-group journaling or questionnaires).

The terms "reflection" or "reflective practice" resonate with me because they imply an ongoing meaningful approach rather than a report out or follow up. Meaningful reflection can and should be woven throughout the entire experience using a variety of methods including metaphoric objects and images, artwork, writing, social interaction, movement, play, as well as dialogue. In this ongoing practice we work to engage learners in meaningful reflection about their experiences so the lessons can be teased out and applied to other aspects of their lives. We want the learner to take ownership of and synthesize the information for themselves and their peers. Reflective dialogue should be an interactive exercise to help learners find meaning in their experiences so they can carry the learning forward. A more participant-centered approach to facilitating reflection can lead us to a wealth of information about a group's progress and ways to improve our teaching that were missed when we took more control of directing the process. Examples of weaving ongoing reflection into lessons are offered throughout the upcoming chapters. Chapter five offers a number of techniques to begin reflective practice from the start and create context and substance around learning experiences that can be threaded throughout. Chapter nine focuses on a broad range of innovative reflection tools and techniques to increase relevancy, depth of understanding, and connection to real life and future learning.

Don't Be Afraid to Try Something New!

Your best teacher is your last mistake. —**Ralph Nader, consumer rights activist**

I always try to encourage facilitators and teachers to welcome the "accidental successes" that arise in teaching and group facilitation and take advantage of teachable moments. Some of my best techniques have come from moments when I didn't have access to the materials I thought I needed, when an activity went awry, or when I had to adapt to a last minute change in setting, time-frame, or the needs of group members. Successful educators experiment in their classrooms, even with lessons that are not fully formed. Don't be discouraged if you present an activity or lesson that flops. That is how we sometimes learn best! You might benefit from asking students for ideas about how it could be improved or changed. Who knows, they may help you develop a brand new lesson or an interesting twist on one of your tried and true lesson plans. Remember you can learn a great deal from activities that don't work quite as you planned. Expect activities to have different outcomes with different groups.

> To teach is to learn twice.
> –**Joseph Joubert, French educator and writer**

Designing and Re-Imagining the Classroom Space

I spend a lot of time moving chairs, and it is worth it! The class or group space has a significant impact on how learners feel, interact, and learn. We have already established that emotional connection to material, movement, novelty, choice, and social interaction are key to enhancing engagement and learning outcomes. It is equally important to set up a physical space or classroom environment that is conducive to this kind of learning. Many of the teachers I work with have moved toward setting up circle, horseshoe, or square configurations of desks in their classroom to allow ease of movement and the ability to go back and forth between

direct instruction and the active review or formative assessment activities described in this book. Many change their classroom setups regularly with help and input from students. Inspired educators include students in the design of the space and spend time creating comfortable, inviting and stimulating spaces where they showcase student work, and students have options on how they situate themselves to work. Having students involved in designing and configuring the room helps them own the space.

A number of elementary and middle schoolteachers have created alternative spaces in their rooms for different purposes. Some have a section of the room fitted with pillows or stuffed chairs for quiet reading time and another space with tables for group project time. Even the smaller, more traditional rooms or spaces can be adjusted to be more conducive to movement and social interaction. Students are highly motivated to move and remove desks and chairs as needed if it means active engagement in the classroom or group room! With a little guidance, they learn to do this quickly and efficiently.

Changing Up the Furniture—Standing Desks

A number of schools both primary and secondary have experimented with alternative seating to improve physical fitness and increase attention with positive results. In the past decade, stability balls have been successfully used by a number of schools in place of desks with numerous reports that they help increase focus and engagement as well as using core muscles and burning calories (Kilbourne, 2009, 2013).

Standing desks are an accepted option in the business world and gaining popularity in schools. Standing desks were utilized by teachers in Minnesota and Wisconsin who were inspired by James Levine of the Mayo Clinic. Levine called for "activity-permissive" classrooms in which alternatives to sedentary activities are encouraged for increased health and fitness. One of the leaders of this charge is Abby Brown and her "Stand Up for Learning" movement. Her initiative utilized the sit/standing work station with stool and footrest for students to use while learning (see resources). Some promising studies on the potential of stand-up desks for increasing physical fitness in both elementary and college classrooms have come out of the University of Minnesota (Reiff, Marlatt, & Dengel, 2012), Grand Valley State University (Kilbourne, 2009, 2013), and Texas A&M (Benden, Wendel, Jeffrey, Hongwei, & Morales, 2012), among others. These studies, which are focused on the obesity epidemic in the U.S., emphasized caloric expenditure and the physical health benefits of standing desks. Though these studies were not designed to study attention, the researchers and teachers involved have shared anecdotally about the increased attention, productivity, and positive on-task behaviors of students using these alternative desks (Blake, Benden, & Wendel, 2012; Deardorff, 2012; Kilbourne, 2013). I look forward to more research on the subject. As leaders in the brain and learning field contend, what is good for the body is good for the brain.

Or No Desks at All!

In the August 14, 2013 issue of *Mindshift*, NPR station KQEDs featured the informative education blog entitled, *To Foster Productivity and Creativity in Class, Ditch the Desks!* Leslie Harris O'Hanlon explored the story of elementary schoolteacher Erin Klein's classroom redesign that involved getting rid of her desks altogether. Instead, she integrated design principles that stores and restaurants use to create appealing spaces in which people want to spend time. With input from her students (i.e., customers), she arranged a variety of places and work surfaces for group projects as well as partner and individual work. She designed the space to feel like a comfortable living room and shared that students treat the classroom with the positive respect they would a living room. There are pages of similar posts online from teachers who have worked with their students to re-imagine classroom space. These inviting and stimulating environments promote brain-based learning principles such as active engagement, emotional connection, social-interaction, movement, and exploration. Research on the brain and learning that emphasizes the need for enriching learning environments (Willis, 2006, 2012) supports these redesign efforts.

Circles: A Key Ingredient for Effective Group Facilitation

My usual configuration for a group conversation or direct instruction is a circle or horseshoe, even if we have to move furniture or move to an alternate space. When a group forms a circle everyone in the group can see everyone else. A physical communal space is created where everyone is on equal footing with equal responsibility. Circular seating or standing arrangements are effective when first meeting a group, giving directions, having a group discussion, or reflecting as a group. As mentioned throughout the book, I often break the group into partners and small groupings, but the circle is the starting and ending point.

When desks or chairs are arranged in a circle or semi-circle, the student who is struggling can't "disappear" into the back of the room. With no space for covert behaviors, students are more able to take responsibility for engaging in prosocial behaviors and participating in class discussion (talking and listening). This approach is common to therapeutic settings for the reasons mentioned above, and it is also beneficial in adult learning situations such as college classrooms or corporate trainings.

Many people are not accustomed to forming circles as a group. Most of society's group forums, such as schools, theatres, churches, and town halls, are traditionally set up in rows. Explain to group members why you are asking them to form circles for activities and dialogue. Forming circles for group interaction might take practice; however, it will eventually become second nature. I notice this transition in groups that I am involved with over longer periods of time. I have worked with a group of students over the past few years that initially needed a good deal of prompting and reinforcement to form a circle. Now, they automatically form one when we start our program without my asking, and they spontaneously circle up without prompting when they recognize they need to share or plan together.

Where Do You Stand?

Where do you place yourself when teaching? Effective educators move around the space as needed to engage the group. Think about how often you place yourself in the front of the room. A change of focus can help the group to feel movement. If you arrange a circle don't facilitate from the center—half of the participants will always have their back to you!

Group Discussions Versus Hand Raising

Hand raising is often the default setting for group discussion in schools even though this is not the method used in the wider world outside of school, and not always the best way to hear from a variety of group members. Even when a group has moved from a formal direct-instruction activity in the classroom into a circular group discussion, I find students (and adults) often raise their hands and look to an adult leader to facilitate the discussion.

I differentiate these types of discussions with students, explaining that though they use hand raising in class at times, today in this group discussion we are not using hand raising. We discuss how, in many life situations, we have to learn to have a group discussion without a facilitator/

teacher calling on group members, and there are active listening and compromise skills involved in this. I challenge them to find ways to make this kind of group discussion work (e.g., learning to wait till others are finished before interrupting, waiting their turn, acknowledging social cues, and speaking up appropriately when they aren't heard).

This is an ongoing practice for both adults and youth. There are different kinds of behaviors needed for different kinds of group situations, and the more we help learners practice a variety of ways to assert themselves and share and listen to others, the more equipped they will be for life outside of school. The partner and small group sharing and even metaphoric sharing activities offered throughout this book give educators alternative methods for facilitating classroom discussion and formative assessment. Learners can practice these skills so they can more effectively use them in large groups. I actually see more participation in open, group discussion than in hand-raising. In traditional hand-raising Q & A, we often only hear from the speedier, more confident processors and not the entire group. There are a few studies that support this observation and even some schools in the U.K. that have completely moved away from hand-raising in the classroom (Dixon, Egendoerfer, & Clements, 2009; Wiliam, 2011).

Attention-Getters

When facilitating interactive activities, it can be helpful to have some "attention-getters" so that you are not yelling at the group when giving directions or facilitating a transition from one activity to the next. Many facilitators and teachers use a raised hand, hand signal, or the overused call and response method, "Clap once if you can hear me, clap twice, and clap three times." I find that many adult groups and middle-schoolers find this patronizing. This type of attention-getter can also be abrupt and feel more like an interruption than a positive call to focus. Like other aspects of facilitation and teaching, mixing up your methods keeps it interesting.

I am continually learning new approaches from the groups with which I work. Whenever I facilitate a large group workshop, I use one of my favorite attention-getters at first and then ask the group to jump in and share the techniques that work for them. This becomes an ongoing, co-creative conversation throughout the workshop.

Including students or group participants in designing or deciding on attention-getters can be helpful in creating ownership. The best methods in my "toolkit" have come from those accidental moments of creativity that arise from a funny situation or comment made in a group. Here are a few favorites:

A Hush Fell Over the Crowd...

The facilitator simply says, "A hush fell over the crowd," and the crowd responds with, "huussshhh." I have been pleasantly surprised to find that middle-schoolers particularly like this one (I think it is the huussshhh sound). I walk around the room and engage a few group members at a time to initiate the progressive attention of the group. While I was facilitating a program in the Pacific Northwest a participant shared, "Around here we would say 'A wave crashed upon the beach' and the group would respond with 'Whoosh.'" My colleague John Lee suggested this technique could be adjusted to many geographic locations or themes (i.e., "A tree fell in the forest." "Timber.").

Call and Response

A simple rhythmic call and response method came from a teacher last summer at the Vermont BEST Institute conference. She vocalizes a "Chi Chi Chi Chi Cha," and her group responds with "Cha Cha Chi Chi Cha."

A Hum

The facilitator starts humming and others join in. I have tried it a few times, and it works quite well. One of the things I like about the humming method is its subtlety as the hum quietly takes over the room.

Visual Cues

Using hand signals can be a nice way to allow participants to finish up their conversation and then redirect their focus without feeling interrupted. My colleague from Japan, Katsumi Namba, uses a simple and appealing visual cue. It is the sign commonly used for applause by people in the deaf community. Hands with outstretched fingers are raised, twisting back and forth in the air.

A Gentle Reminder

Polly Chandler shared this comment on my blog:

For someone with a soft voice, getting a large group's attention has been a lifelong challenge. Many attention-getting strategies are just annoying and noisy. I sometimes do a little tapping on the wall in a drumming pattern. I have also found that walking around the room and letting people know they have 1 minute helps.

Freeze Like Trees

A participant in an environmental educators workshop shared that, in her work with students of all ages, she calls out "Freeze like trees." Group members "freeze" in place in a "tree position" of their choice. I have tried this with adults and college students who were quite creative in their poses.

More Cowbell

Sound-makers can be effective attention-getters. I have a train whistle, vintage bike bell, and cowbells in my facilitation toolkit. The downside is I have to keep track of them, and sometimes, they seem to lose their effectiveness (probably because they are so facilitator directed). Here in Vermont, the cowbell is still a favorite.

Let Your Group Show You!

Marci Charles shared that the youth in her Team BRIDGES Experiential Learning Center program, in Memphis, TN, use stepping or step dance as a call and response attention-grabber. Step dance is a form of percussive dance in which people use their body as an instrument, producing sounds and rhythms through a mixture of footsteps, handclaps, and vocalization. Marci's first experience using this as an attention-getter occurred when she told a large group of high school students that they were going to need some way to get everyone's attention. Immediately a student responded with a couple of knee slaps and stomps and others followed. Throughout the day, the group built upon "the step" as they used it as their attention getter. It became a collective celebration of their day together. There was a great deal of buy-in because the students created it, owned it, and it combined visual, auditory, and kinesthetic engagement.

Circle 'Em Up

Purpose/Focus: organizing groups, attention getter, playful learning, active engagement, collaboration, movement, social-emotional learning, following directions

We've established that circles are a key ingredient for effective group facilitation. Often we need circles of different sizes for different group situations. My colleague Don Morse invites groups to form circles of different sizes using what he calls "Chowder, Chicken, Flamingo." The first day I experienced this method with Don, I was amazed at how a "too cool for school" group of 8th graders bought into this activity.

Facilitation Suggestions:

- Show the group that when the facilitator says "Chowda" (New England for Chowder), group members form a tight knit circle that looks like a bowl of chowder.
- When the facilitator says "Chicken," group members take one step back to make a slightly larger circle and put their hands on their hips with elbows out like chicken wings.

- Finally, when the facilitator says "Flamingo," group members take three steps back, spread their arms, and bend one leg, making a large, spread out circle.
- Have participants practice each response so they will know what size circle to make when you call out "Flamingo."

Facilitators from other parts of the U.S. have shared regional variations of this activity such as the Southern "Gumbo, Chicken and Flamingo" and the Pacific Western "Fog, Redwood, Ocean."

Innovative Strategies for Dividing Groups

Educators and facilitators often need to divide large groups into smaller groups or teams for activities or discussion. I use a number of simple techniques to facilitate this process:

Playing Cards and Dominos

Simple props can be used for organizing small groupings and teams. They work well because the group sees them as random pairings initiated by "luck of the draw," which increases buy-in and willingness to work with people outside their usual social group. Chapter six offers techniques using simple props such as playing cards, dominos and matching key blanks for dividing a group into pairs.

Sections of a Puzzle

When you want to create mixed groupings and add a little fun, try taking apart a puzzle and grouping the team by puzzle section. For example, I use children's puzzles of dinosaurs and sea creatures (found in the toy section of my local bookstore). Just take the pieces that make up the tyrannosaurus, the octopus and the whale and mix them together. As group members arrive, hand them each a piece and ask them to find the group that their piece fits. Putting the puzzle together becomes a mini team-building exercise as well as a way to get participants interacting with those they might not know well. Another idea is to use a puzzle that connects with a subject you are exploring in order to create context.

Birthday Seasons

When four groups are needed, try dividing the participants by the seasons in which they were born. If the birthdays don't make for an even distribution, even things out by asking a few people to move over to their favorite season or the season of their favorite sport or activity.

Arm Cross/Hand Cross

Simply ask participants to cross their arms over their chests. Those people with their left arms on top form one group, and those with their right arms on top form the other. This also works with clasping hands.

Grouping by Sound

Todd Loe, a Driver's Education Teacher at Milwaukee Public Schools, shared his strategy for grouping up students in his driver's education classes by sounds. He calls it his "six dogs" strategy. He made various cards using funny pictures of dogs and labeled them with sounds: "sniff," "bark," "snore," "howl," "growl," and "arf." As students enter the room for class, he gives each a card and asks them to group up using their dog sound.

Which One?—Partner Decision Making Method for Dividing into Teams

Purpose/Focus: dividing groups, conversation starter, playful learning, social-emotional learning, 21st century skills, communication, compromise, collaboration, decision-making, conflict resolution

My "go to" activity for dividing groups into teams comes from my time teaching with Dave Lockett in Stevens Point, Wisconsin School District's Experiential Education program. After becoming bored with the old count off "1-2" method, I experimented with this fun and engaging way to divide a group and, at the same time, help participant's practice communication, decision-making, consensus, conflict-resolution and compromise with peers. This method honors participants' needs for a sense of choice and control over who is going to be there partner. It can also help you subtly separate participants when needed to mitigate behavioral issues and increase diverse interaction.

Facilitation Suggestions

- Have everyone find a partner. (With student groups, participants most often pick a buddy. This method of dividing honors their need to partner with a friend because it gives the friends something to do together before they split up.).
- If there is an uneven number, step in as a partner.
- Present a hypothetical situation to the group with two choices:
  It's a beautiful summer day, and you and your partner would like to go paddle on the lake. You go to the dock and find there is only a single-person canoe and a kayak available. Who is going to take which?

Or: *You and your partner go to an ice cream stand. It is late in the day, and there is only one chocolate cone and one cookie dough cone left. Who will choose the cookie dough and who will choose the chocolate?*

Or: Use a scenario that involves something you are studying in class.

- The partners have to compromise about who gets which choice in the scenario.
- Once the partners decide, the canoes become one team and the kayaks become the other team (cookie doughs become one team, chocolates the other).
- If multiple smaller groups are needed, repeat the exercise with new partners and different choices until the desired group size is reached.

Over the years, I have found that groups of all ages buy into this method of dividing teams without "pre-arranging" themselves the way participants often do with the old 1-2 count off method. This method honors their need to pair with friend before being parted.

It can be surprising how intense the discussions and negotiations around these imaginary decisions can be! Younger participants can practice and reinforce conflict resolution, positive decision-making, compromise, and consensus. Adult participants enjoy the sense of connection and humor from sharing and making decisions about these hypothetical scenarios.

Resources/References: *Tips & Tools* by Jennifer Stanchfield; Middleton High School Experiential Education Curriculum, 2001: J. Stanchfield.

Facilitating Healthy Competition

In chapter three we explored the power of play to engage and motivate students and to teach important social-emotional skills. Learning how to compete in a positive, pro-social way is one of the valuable lessons that can be learned from play experiences.

When I worked with boys in a juvenile correction program, I stopped using competitive games altogether because the boys had great difficulty maintaining appropriate behavior during competitive games like basketball. What I realized soon after creating this moratorium on competition was that I was not helping them learn healthy skills and pro-social behaviors. They needed to learn how to compete in a healthy way. The staff and I brought back competition by allowing them to play basketball but adding the rule that, as soon as they scored, they had to switch teams with another player.

This shifted their focus from outcome or score focus to process and the joy of the game. Since then I have used this team switching method in a variety of games and activities.

After a few turns, simply call out random topics such as "everyone who is wearing blue" or "anyone whose birthday is in the winter," and have those participants switch teams with someone else. Group participants actually seem to enjoy the regular switching. It adds a little fun and chaos and pretty soon groups become unaware of the score—focusing instead on the joy of playing.

Note: I rarely keep score in competitive games (one exception being the Telegraph Probability activity, page 120). Most groups don't seem to miss it (with the exception of some highly competitive teacher groups during a staff professional development day).

Cultivating Creativity

Play Dough Pictionary (page 110) is one of my favorite community-building and academic-review activities. One of the reasons it stays at the top of my list is that it has led to many meaningful discussions about creativity. After playing the game, I ask participants to share one word to describe the activity. The words "imagination" and "creativity" come up regularly. Interesting responses occur when I ask the group, "How many of you consider yourselves creative?" In a 3rd grade class, almost everyone raises their hand. When I ask the same question of students in 8th grade or high school or adults, very few do.

Children will give many examples of using creativity whether it is in making up a game, working out a conflict with a friend, or building a fort. Older students or adults will often only refer to artistic ability. When asked why they didn't raise their hands, they will say something like, "Well, I am not a good artist."

This is a great opportunity to explore the true definition of creativity with older group members and help them reflect on examples in their day-to-day life when they ARE creative. These examples of creativity have a wide range (e.g., figuring out a way to effectively work out a conflict with a peer or family member, finding a way to fix something when they don't have the "right" tools, making an argument to persuade their parents to see something their way, writing a paper in English, working out a system for

Creativity
The ability to make new things or think of new ideas. The ability to produce something new through imaginative skill, whether a new solution to a problem, a new method or device, or a new artistic object or form. The term generally refers to a richness of ideas and originality of thinking.
Synonyms: cleverness, ingenuity, inventiveness, originality, imagination, innovation, Cultivating Creativity
–Merriam Webster Online Dictionary

Dr. Jonathan Plucker, a leader in the study of creativity and learning from Indiana University, defines creativity as "the production of original, useful things" (Whirty, 2002).

keeping group members accountable in a group project). I try to help the learner realize that everyone has the ability to be creative, and it is a skill that can be practiced.

One of my strongest arguments for promoting experiential approaches is that they promote creativity and creative problem-solving skills. Creativity, in my view, is one of the most important life skills. Proponents of the 21st century skills movement would argue it is one of the top skills necessary to thrive in the 21st century workplace.

Fifty years ago, E. Paul Torrance pioneered the Torrance Test of Creative Thinking. In longitudinal studies that followed, there was a strong correlation between those with high scores and their success and accomplishments as adults. When Jonathan Plucker reanalyzed the data in 1999, the correlation to lifetime creative accomplishment was more than three times stronger for childhood creativity than childhood IQ (Bronson & Merryman, 2010).

As educators we can cultivate creativity in learners by engaging them in insightful problem-solving tasks and giving them opportunities to experiment, collaborate, explore, make metaphors, take ownership over their learning, take risks in learning, and by asking more questions than giving answers. This book focuses on a number of activities and resources to help educators structure opportunities for inquiry, exploration, and creating meaning from metaphor. Educators also cultivate creativity when they are willing to go with the flow of lessons being created by their students and take advantage of unplanned opportunities for learning that arise. When your group takes an activity or discussion somewhere unexpected, be flexible if appropriate. Being open to these opportunities and knowing how to guide them is one of the qualities that makes for a successful educator. Nurturing creativity in learners is what experiential education is all about.

Chapter 5

Strong Beginnings

Engaging Learners From the Moment They Walk in the Door

Snapshot: It is August 28th, the day before school starts. Middle school science teacher Emily Case is ready. She has the cards and objects she collected over the summer to use for pairing and reflection activities. On the whiteboard, she has an inspiring quote on new beginnings to start the year. The front of the room has been arranged for an engaging entry task to welcome students as they enter her classroom. Emily has a collection of postcards ready for a "get to know you" activity that will also create context for upcoming lessons. She emails photos of these preparations to colleagues, sharing with them her excitement and readiness to start off a new school year with her students.

Starting Off With Style

The beginning is the most important part of the work. –**Plato**

Think back to your most memorable learning experiences, the courses you enjoyed, and the teachers/facilitators you found most effective. How did they start their classes or workshop sessions? How did they begin the school year? How did they greet you when you came into their classroom? Chances are they intentionally planned to draw you into the learning experience.

Many educators do this instinctively. We consider ourselves hosts to the participants or students joining us for a learning experience. For years, I have been using objects, postcards, or quotes and entry tasks such as writing prompts, reflective partner conversations, or a daily puzzle as a way to welcome and engage group members from the moment they walk in the door. These methods involve the group in a meaningful learning experience and provide them with a sense of belonging, purpose, and connection while at the same time allowing me to take care

Note: There is a great deal of cross referencing between chapter five and chapter nine connecting the meaningful activities that start off a program and the reflective activities used to create connections. It is my belief that meaningful reflection begins the moment learners walk through the door.

of administrative duties such as adjusting room setup or materials, taking attendance, or collecting homework.

In the U.S., people are often late to workshops and the entry time from the first person arriving to the last can extend for 20 minutes or more (though I found this was not the case in my recent travels to Japan!). Using an entry activity leaves me free to greet the late arrivals, making me feel like a better host. Research in the field of educational neuroscience confirms this strategy can be a valuable evidence-informed practice that transfers to day-to-day classroom teaching and training programs. Intentionally creating an engaging and novel start to the session has many benefits to learners.

Proponents of brain-based learning emphasize that events that happen the first time learners are exposed to information greatly impact their ability to accurately retain the information. John Medina, author of *Brain Rules* states: "If you are trying to get information across to someone, your ability to create a compelling introduction may be the most important single factor in the later success of your mission" (2008, p. 116). The first few minutes of the learning experience or lesson are a key time to hook and engage learners. Educational psychologists have demonstrated that people remember most the first few minutes and the last few minutes of a learning experience (Sousa, 2006; Willis, 2010). Psychologists call this the primacy-recency effect. This underscores the importance of facilitating an engaging opening activity as well as providing some reflective prompt to tie it all together or "bookend" a learning experience. This suggests it also might make sense to create as many introductory and closing moments as possible in your teaching and group facilitation.

As discussed in chapter one, the reticular activating system (RAS)—the brain's sensory intake filter through which all sensory input must pass—is very receptive to novelty and change, so it responds to pleasure and sensory input that arouses curiosity (Willis, 2010). Novelty and curiosity-evoking events alert the RAS system to pay attention because there is something new or different that warrants further evaluation (Aamodt & Wang, 2011).

Find a Hook!

Engage learners from the moment they walk in the door. Involve participants in an activity that helps them transition positively into the learning environment rather than using these precious moments to focus solely on taking attendance, collecting homework, or other administrative duties. This can be a time for them to make positive connections with their peers, explore or review the academic material at hand, create context around a lesson, and most importantly, shift their focus to the here and now.

Coming together for the first time can feel awkward for some participants. A novel activity engages learners right away, drawing them into a positive experience and creating buy-in. Learners of all ages come to the classroom with things on their mind—a stressful commute, a rough morning at home, a negative social interaction in the hallway before class, or the exam coming up next period. A reflective, focusing activity can help learners transition into the classroom or workshop space and shift their focus to the present and the lesson at hand.

People are drawn in by colorful, novel, or intriguing objects whether they are tools, old toys, or items found in nature. People often share and reflect more readily when they can attach their thoughts and feelings to an item that can be touched and shown to a group during discussion, or described in reflective writing. As discussed in chapter one, brain research suggests that using metaphors, pictures and symbols helps cement lessons and transfer learning to everyday life and future learning (Willis, 2006, 2010). Throughout this chapter and chapter nine we will explore many metaphoric methods used for creating strong beginnings and facilitating meaningful reflection throughout lessons and programs. Here are a few of my favorite approaches for engaging groups from the moment they walk in the door.

Images/Postcards

Purpose/Focus: active engagement, hook, "first five" or "do now," conversation starter, context setting, discussion or writing prompt, reflection, community building, figurative language, metaphor, emotional connection to learning, creativity/imagination, differentiation, multiple pathways to learning, choice, control and ownership, group norms

Material: postcards, pictures from magazines or other images

I often use my "Pick-A-Postcard" collection as a way to draw learners into a classroom or group experience and create context around an upcoming discussion or lesson. It works well as a get-to-know-you and/or reflection activity on the first day of school, to start a new program, or as a transitional activity to welcome groups after a break between workshop sessions or school vacations. Images are an engaging way to explore the use of metaphor, figurative language, and perspective. They also make for great creative writing prompts.

Facilitation Suggestions

- Spread out the postcards/images so they are accessible to all group members.
- As participants enter the room, ask them to choose a card centered on a reflective, goal-setting, or conversation-starting topic:
  - › Choose a postcard that you would send to a friend to describe your summer break.
  - › What image describes your experiences in this course so far?
  - › Choose an image that represents a goal that you have for the program or your expectations for the day.
  - › What image represents a unique perspective or strength you bring to the group, class, or program?
- Pose a question related to the group's purpose for coming together. For example, when leading a program with a group of teachers you might use one of the following prompts or questions:
  - › Choose an image that represents a reason you are drawn to teaching.
  - › What card represents a quality you think is most important in a leader?
  - › What image represents a strength you bring to your work?
  - › Choose an image that represents a hope you have for the upcoming school year.
- As participants of an ongoing group or at the end of a course enter the room, have them choose a card describing their experience to could send home.

Facilitation Notes

Having the cards available during pre-group gathering time or as an entry activity can help jumpstart conversations as group members get to know each other and set the tone for the program. Choosing a card gives people something to focus on as they transition into the workshop or classroom space. Depending on the group, participants can share their choice with a partner, or write about their choice. When group members are comfortable with each other, have them share with the whole group. For using postcards/images in reflection, see chapter nine.

I like to blend this activity with an interactive partner sharing activity such as Commonalities Mingle (see page 97) or Concentric Circles (see page 86).

Academic Variations

Using images is not only great for building community or practicing social and emotional skills such as communication, insight, and reflection, but it can be a way to introduce, reinforce, or review academic material:

- A language arts teacher who wants to reinforce the idea of "theme" might lay out postcards and ask students to choose a card they are drawn to and then identify their interpretation of the card's theme.
- In a creative writing course, students might be asked to write a story about their card or choose a card that describes them and use figurative language to explain why.
- A social studies or science teacher might choose images around a particular subject area and ask students to choose an image that resonates or speaks to them as a way to create interest.

Reference/Resource: Variations of object and image reflection activities are described in more detail in chapter eight and nine. Other books include *A Teachable Moment* by Cain, Cummings, and Stanchfield; *Tips & Tools* by Stanchfield; and *The Chiji Guidebook* by Cavert and Simpson. Make your own image or postcard kit with found items. Pick-A-Postcard kits are available at www.experientialtools.com.

Toolbox, Object, or Miniature Metaphor Charms

Purpose/Focus: active engagement, hook, context setting, discussion or writing prompt, "first five" or "do now," conversation starter, reflection, community building, figurative language, metaphor, emotional connection to learning, differentiation, 21st century skills, creativity/imagination, choice, control and ownership, multiple pathways to learning, group norms

Material: a collection of objects

I use metaphoric objects in much the same way that I use postcards and images. My old toolbox filled with a collection of found objects (i.e., antique keys, old lock, vintage camera, joke glasses, white out, a ruler) creates a great deal of intrigue in the classroom. I developed the "Miniature Metaphors Kit" (see resources) in response to an Outward Bound instructor who shared that he was looking for small reflective tools he could take into the field.

Facilitation Suggestions

- Using tangible objects is especially effective when introducing the subject of roles or responsibilities in group work or to celebrate strengths or accomplishments:
  - As participants enter the room, have them choose an object that represents their role in the project. Or, have them choose the unique perspective or strength they bring to the group.
  - Ask participants to join their work groups and share their objects.

This can be a great way to help participants address some of the difficulties around communication and teamwork that come up during group projects and turn these challenges into teachable moments around roles and responsibilities, accountability, and establishing positive group norms.
- For language arts classes, the tangible objects can start a reflective writing assignment or discussion about descriptive or figurative language around personal strengths or goals.

The Most Important "Tool in Your Toolbox" Variation

- Fill your toolbox with a variety of household tools: hammers, tape measures, a whisk, oil can, glue, extension cord, etc.
- As participants enter the room, ask them to choose a tool that represents the most important "tool" to have in one's toolkit. Explain that this could be a teaching or learning technique, a philosophical view, attitude, etc. (I find this creates context for workshops on teaching and group facilitation and jumpstarts thinking on a subject.)
- It can be effective to come back to this tool choice at the end of the training to reflect upon new perspectives gained or lessons learned.

I find that the more energy I put into learning about the group and what they do before the program starts, the better the outcome. A more structured program tailored to that specific group results in the group digging deeper into the activities and getting more to take home with them.

Workshop participant responses:

"I chose the Allen wrench set, because facilitating is often like picking a wrench and putting it into the hole only to find out you have the wrong size, so you keep trying until something fits, gains leverage, and turns for the better."

"In my mind the most important key to good facilitation is being flexible, so I chose the elastic. I come prepared with an agenda and goals for the day, but I have to adjust to the changing needs of my groups and welcome the unexpected opportunities for learning that arise."

"I chose the matches: To me they are a symbol of using activities to ignite the spark of learning in participants."

"This battery represents the importance of energy. The level of energy and passion a facilitator puts into a program will play a big role in the outcomes experienced in the group. If I am bored with the activities I am facilitating, then why should I expect the group to be excited about them?

Reference/Resource: Many variations of object and image reflection activities are described in more detail in chapter nine. Other books include *A Teachable Moment* by Cain et al., *Tips & Tools* by Stanchfield, as well as *The Chiji Guidebook* by Cavert and Simpson. You can easily make your own reflection treasure chest with found objects, old toys, or yard sale items. Miniature Metaphor kits are available at www.experientialtools.com.

Quotes

Purpose/Focus: active engagement, hook, introducing a topic, "first five" or "do now," context setting, conversation starter, reflection, discussion or writing prompt, emotional connection to learning, differentiation, formative assessment, community building, transition

Material: a collection of quotes

Quotes are another effective transitional, introductory, or reflective activity. They are an engaging way to explore an academic subject, a discussion topic, or a lecture. Discussing quotes can help group members connect with each other or spark creative thought around a subject.

Facilitation Suggestions

- When facilitating workshops for educators, I often start out the program by displaying my collection of quote cards with themes around leadership, teaching, and learning.
- As group members arrive for the program, I ask them to choose a quote that resonates with them. (Group members often initiate conversations with each other during this pre-workshop time as they gather at the quote table.)
- I discovered that leaving some blank cards out with the quotes can increase participation from group members who would rather make up or share their own quote. This idea came as one of those "accidental successes" in facilitation when I mistakenly left some blanks in a stack of quotes. Participants (especially those who needed a sense of ownership in a workshop they were forced to attend) liked having the opportunity to create their own.

- Depending on the program, participants can later reflect upon this individually, write about it, or discuss it with a partner or the whole group (see pages 86-93, 95-96 for partner dialogue ideas).
- Quotes can also be used again later in group process as a reflective tool (see chapter nine).

Academic Content/Creating Context Variation

Recently, I found success in using quotes to introduce academic material. An 8th grade social studies class was about to begin their unit on the American Colonial period and the Revolutionary War. I worked with their social studies teacher to find quotes from people of that era that the students would be studying and quotes describing the events of the era. We printed them on card stock. This exercise not only piqued their interest about the subject and the upcoming material, it also increased the depth of classroom discussions throughout that unit.

- On the first day of the unit, display the quotes and ask students to choose one that resonates with them.
- Ask participants to share the quote and why they chose it with a partner or small group.

Outcomes/Reflections

- The teachers were able to make a formative assessment as they heard students discuss their choice and their understanding or misunderstandings about that period in history.
- Students seemed to want to know more about the people behind the quotes they chose. Many were curious enough about their quote's author to take the initiative to read ahead in their social studies book or research information about them online.
- When events and people related to the quotes came up in reading or lecture, students recognized the connections and were able to better relate to the events or person they had discovered through this activity.

Facilitation Notes

I have also used quotes in language arts to introduce a unit on theatre and in guidance group to frontload a discussion on empathy. I saw many of the quotes from all three classes in the front of student's binders or taped to the inside of their locker!

The process of collecting quotes for use with your groups can be a rewarding reflective task for an educator. This reflective exercise has sparked my own thoughts around subjects as I planned and prepared for a lesson. There are some great quote books and websites. I especially like to seek out vintage quote books in used bookstores. Enjoy your search!

Reference/Resource: Make your own, there are some great quote books and websites. "Quotables" cards are available through experientialtools.com.

"Conversation Starter" Pin-Back Buttons

Purpose/Focus: active engagement, hook, playful learning, context setting, reflection, emotional connection to learning, community building, discussion prompt, social-emotional learning, choice, control and ownership, self-expression

Material: a collection of pin-back buttons with playful quips and sayings (see resources)

Conversation-starter buttons are a fun and easy way to transition into a classroom or workshop space. They can be used with middle and high school students as a way to check in on progress around group projects. This is also a tried-and-true activity for engaging educators or college leaders in laughter and dialogue during a professional development session. The buttons bring humor into group settings where individuals are reluctant to participate, helping them buy-in to the group process.

Facilitation Suggestions

- Lay out the buttons on a table. As participants enter the room, ask them to choose one that represents their mood or attitude.
- Having the buttons available during pre-group gathering time or as an entry activity can generate conversations, help group members get to know each other, and set the tone for the program.
- Generally participants spontaneously share about their choices as they meet each other.
- Instead of asking participants to share in a "sharing circle," I often integrate sharing about their buttons into partner dialogue activities such as Handshake Mingle, Concentric Circles, Commonalities Mingle, or Trade and Share (see chapter six).

- For groups who are working on a task or project together, the buttons can be used to jumpstart reflection on their roles, responsibilities, goals and challenges with the group process (see chapter nine).
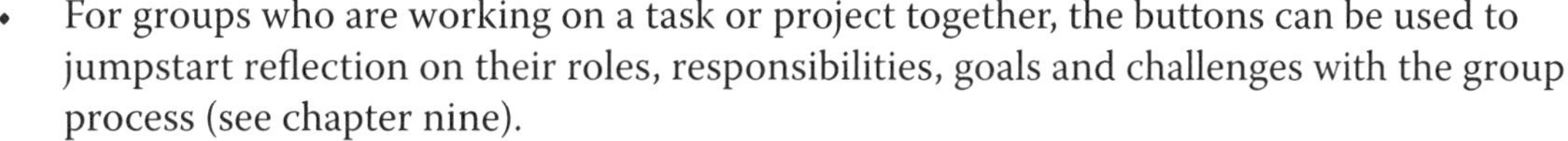

Reference/Resource: You can make your own buttons with a button maker. "Conversation Starter" button kits are available through experientialtools.com.

Computer Keyboard Keys

Purpose/Focus: active engagement, hook, context setting, "first five" or "do now," reflection, community building, metaphor, descriptive language, discussion/writing prompt, social-emotional learning, creativity/imagination, multiple pathways to learning

Material: keys from a computer keyboard

Andy La Pointe, career development specialist and challenge course facilitator at You Inc. (a therapeutic youth residential treatment and school program in Massachusetts), was inspired by our workshop on processing/reflection tools. When I arrived at his site for another workshop, he had a bag filled with the keys of a keyboard that he had found in a recycling bin. We used them that day, and I was impressed with the conversation the keys initiated and the connections group members made to various keys. It can be surprising where conversations can go with such a simple tool.

Facilitation Suggestions

- Display the keyboard keys on a table. As participants enter the room, ask them to choose a key that represents a goal they are working on for the class, project, or school year.
- Instead of asking participants to share about their key choice in a "sharing circle," I often integrate the sharing into partner dialogue activities such as Handshake Mingle, Concentric Circles, Commonalities Mingle or Trade and Share (see chapter six).
- For groups that are working on a task or project together, use the keys to initiate reflection on their roles and responsibilities, goals and challenges with the group process (see page 171).

Academic Variation Examples

I have used the keyboard keys for introductory or transitional activities as well as a processing tool. On the first Monday in January after holiday break, I asked middle school students as they entered the classroom to pick a key that represented their new year's resolution. The way the students used the keys to represent their hopes and goals was creative and memorable.

One inspired teacher used this activity with colleagues and students at her alternative school. She used it at the end of some team-building work with students, asking the question, "What was the hardest part about your work today?" She also used it at a staff training, asking the question, "How would you describe your teaching style?" And, having never met her students before, she planned to use the activity as an icebreaker to open class. The lesson was on Excel® and how it relates to a technology education project (truss bridges). The main point she wanted to convey was that Excel would make their lives easier by simplifying the math involved. She was thinking about using the question, "Which key describes how you feel about math?" or "Describe yourself as a math student."

> "I chose the escape key because I know I need to make some better choices about who I hang around with or sit next to in class. I need to "escape" from my friends and distractions so I can get my work done and not get into trouble."
>
> "I chose the home key because I would like to try and get along better with my stepbrother."
>
> "I chose the question mark key because I know I need to ask more questions and get help during homework club time so I can improve my grades."

Facilitation Notes

Recently, I acquired keyboards from Macintosh and PC desktops and laptops. I mix them together when using them with groups. This leads to some interesting and humorous conversations about learning and personality styles.

As with many of my tools, I used these the first time without having a clear picture as to how it would work. There have been many pleasant surprises. I encourage you to take a new look at your recycle bin and see what you might come up with!

Studies on learning emphasize the value of strong beginnings and endings with a hook to engage learners at the beginning of a lesson and a good closing activity to "tie it all together." I often use the same activity for both purposes to "bookend" the lesson or experience.

Resources/References: The *Inspired Educator Blog* by Stanchfield www.experientialtools.com. As this activity illustrates, you (or your students) can easily create your own tools. Computer stores and schools often recycle keyboards. Thank you to Andy LaPointe for your inspiration.

Journaling or Reflective Writing Prompts

Purpose/Focus: context setting, "first five" or "do now," reflection, review, transition, 21st century skills, communication, self-expression, multiple pathways to learning, differentiation, formative assessment

Material: paper and writing implements

Engaging students in a short writing exercise at the beginning of class helps them transition into the learning space. Heidi Pauer, a high school language arts teacher from Bow, New Hampshire, starts each of her classes with "First Tens" (the first ten minutes of class), using reflective writing prompts. She writes a prompt from *The Me I See, 2E* (see resources) on the board.

Heidi Pauer's Suggestions for First Tens Writing Prompts

- As students come into the classroom, let them know that they are going to write for the first ten minutes to warm up for class and to transition from their previous class.
- Give them the choice to write about the prompt or anything else. (In the busy life of teenagers, they often don't have quiet time to think about themselves in a non-threatening way and to process what they are becoming.)
- When students are done with first tens, invite them to share by reading all or part of what they wrote (this is always a personal choice by the students). For example, the group could talk about how, as writers, it would be interesting to see how answers to the same questions might change throughout our lives.

Pauer's Outcome Notes

Heidi writes with her students and shares her thoughts as well. This helps build community and trust in the classroom, which is essential to the success of ANY class for that matter. At the end

of the semester, Heidi asked students what they liked best about the writing courses she taught. Most of the students said the first tens' writing prompts were their favorite because they couldn't be wrong; they found them to be a nice transition from "regular" classes; and they had the opportunity for self-reflection as well as learning new things about their classmates.

Variations

In my own ongoing work in primary and secondary classrooms and with adults in professional development programs, I regularly use reflective writing prompts as an entry task to get a group thinking about a subject before delving into a lesson on the topic. For example, to kick off a lesson on poetry with 7th grade students, I used the following questions to help students explore their prior experiences with poetry through music. I found it was helpful to allow them to use their iPods in class to look at their playlists or to assign them the task ahead of time as homework to think about prior to this exercise.

- Think of a song you love and remember the words to.
- What song is it? Who performed it? Describe the song.
- What is it you like about this song?
- Why did it make such an impression on you? Why do you remember the verses?

After students completed the reflective writing, they shared in small groups. We found this entry task helped them buy into the poetry lesson and engage in a larger group discussion.

Resources/References: *The Me I See, 2E: Learning Through Writing and Reflection* by Wood N Barnes Collective and *Writing and Experiential Education* by Rapparlie.

Daily Riddle or Brainteaser

Purpose/Focus: active engagement, hook, "first five" or "do now," transition, conversation starter, creativity, critical thinking, problem-solving, playful learning

Material: a brainteaser or riddle

Facilitation Suggestions

- Post a daily brainteaser or riddle for students to work on as they enter the room.

- Teachers can turn these puzzles or riddles into an ongoing contest with students receiving points for the first answer, the right answer, etc. I have had teachers report that students come to class early so they can start solving the daily riddle.
- Eventually students might bring in riddles or problems of their own to post.
- Riddles can be developed around specific subjects, current events, or other relevant topics.

Resources/References: There are lots of resources for this, including *Mind Trap* by Mind Trap Games, Inc.

Question or Conversation Starter of the Day

Purpose/Focus: active engagement, hook, "first five" or "do now," conversation starter, context setting, community building, reflection, social-emotional learning

Material: a question for the day (e.g., *Chat Pack* from Questionguys.com, *Table Topics* from tabletopics.com), paper and writing implements

Questions can be about curricular content or the previous night's reading. They can be reflective, designed to get students thinking about the school climate, current events, or their personal strengths or goals.

Facilitation Suggestions

- Ask students to quietly write or doodle about the question or look up answers and related information online. Students could also share their thoughts or answers with partners during the first few minutes of class.
- A great way to pair up students for these partner conversations is to use the Domino activity or match game partners mentioned in chapter four.

Variation

Display a selection of questions on a table near the classroom door. As students enter, ask them to select a question they want to answer or think about. Then use active pair share and other activities to initiate conversation about the questions

Emily Case, a middle school teacher from Hadley, Massachusetts, uses these questions: Which phase of the moon would you be and why? Who would you rather eat lunch with ____ or ____? (Insert literary or historical character.)

Group Sculpture for Academic Review or Reflection

Purpose/Focus: active engagement, hook, "first five" or "do now," reflection, academic/content review, formative assessment, differentiation, multiple pathways to learning, metaphor/imagery, collaboration, community building, social-emotional learning, 21st century skills, creativity and imagination, innovation, problem-solving

Material: items for scuptures (i.e., play dough, dominos, puzzle pieces)

My colleague, Jessica Hammond, and I experimented with this idea, hoping to engage students as they entered her science classroom. We placed cans of play dough in the center of each desk grouping or "pod" (4 to 5 students). Students were asked to work together to create a sculpture that showed their understanding of one of the force and motion concepts they were studying (i.e., acceleration, centripetal force, etc.). Students loved the activity which drew them into her lesson for the day and acted as a formative assessment. Since that time, I have used it in other subject areas and as a reflective activity in group-building.

Facilitation Suggestions

- This works well as an entry task or formative assessment for an ongoing project, building upon prior knowledge in each session.
- Provide props such as play dough, dominos, puzzle pieces (whatever is handy in your toolkit) and ask groups of participants to work together and create a sculpture representing their understanding of a concept.
- As a reflection activity, participants can sculpt a moment or concept that stood out for them in a previous session or a key take-away from a previous session that they might use again that day.

Musical Connection

Purpose/Focus: hook, active engagement, attention getter, transition, context setting, introducing a topic, multiple pathways to learning, differentiation, emotional connection to learning

Material: music on iPod or CD, musical instrument(s)

Music is a powerful tool for drawing learners in and creating connections to material or to community. Playing a song related to the academic material as students/participants enter the room promotes curiosity and focus (Willis, 2011b).

Facilitation Suggestions

- A social studies teacher introducing a colonial history or civil war unit might play a song from that period or event to spark conversation and reflection around the subject.
- A language arts teacher might use songs to engage students in reflecting about poetry or storytelling.

Graffiti Wall

Purpose/Focus: active engagement, hook, emotional connection to content, "first five" or "do now," transition, conversation starter, community building, empathy, discussion/writing prompt, reflection, collaboration, differentiation, multiple pathways to learning, creativity/imagination, group norms

Material: butcher or flip-chart paper, masking tape, markers, crayons, etc.

Kasey Errico shared with me that she and her colleagues use this activity to encourage creative thinking prior to staff meetings. I find this effective with teacher groups, students, and leadership programs for reflection, to initiate conversations about a subject, or to review material.

Facilitation Suggestions

- Tape lengths of paper on the walls or a table top.
- As participants enter the room, hand them magic markers or crayons to represent a can of spray paint and ask them to imagine that they are graffiti artists who have been invited to express their thoughts on the topic at hand using words and symbols.
- As with graffiti, others may add to and/or comment on these musings.
- The activity can be done silently to increase focus.
- For longer courses or workshops, participants can add graffiti throughout their time together (see chapter nine).

Outcomes/Reflections

I like the blend of self-reflection and group collaboration found with this activity. Often people will more readily express their thoughts through artwork than verbal means. It is non-intimidating to those who don't consider themselves "artists" because they can blend words and images. Group members can choose their level of participation and learn as much from contributing as they do from observing other's posts. The graffiti becomes an "artifact" of the group's thoughts and experiences and can generate meaningful dialogue and reflection. Doing this silently (a challenge with middle-schoolers), increased the reflective aspect and focus during the activity. See chapter nine for specific examples of using this and other creative expression activities for reflection.

Summary

These are just a few of many techniques that can help you make the most of your teaching time, initiate reflective practice, help students take ownership over their learning experiences, and increase focus, attention, and retention from the very start of a learning experience.

At the very least, take the time to greet your students/participants at the door (by name, if possible). There have been a number of interesting studies about the positive effects on student outcomes in classrooms where teachers and professors greet each student at the door as they enter the classroom (Allday & Pakurar, 2007; Weinstein, Laverghetta, Alexander, & Steward, 2009). These studies, involving both college students and middle school students, found that on-task behavior and test scores increased in classrooms where educators took the time to say hello to each student.

It is our attitude at the beginning of a difficult task, which, more than anything else, will affect its successful outcome. - **William James**

There is great opportunity for positively influencing learning outcomes by intentionally designing a strong beginning or "hook" for your lesson or group experience. Whether it is the start of a new program, the first day of the school year, the start of a single classroom session, or the beginning of a unit, the tone we set in the classroom and the frame and context we set for a lesson can greatly impact student engagement and retention.

Chapter 6

Get Them Moving, Talking, Reflecting and Keep Them Engaged

Multiple Pathways to Learning

Snapshot: As the staff of the mental health agency arrives at their annual all-staff training and retreat, they find tables arranged around the perimeter of the room instead of the usual PowerPoint® and rows of chairs. A wall is covered with blank paper and markers and various props are spread around the room. As participants enter the room they are asked to choose a quote from those scattered on a table that resonates with them. The participants are drawn in by the quotes and intrigued by the unique setup. As people continue to settle in, the facilitator walks around the room with a box of dominos and asks everyone to choose one. The group's curiosity is piqued and everyone is waiting to see what will happen next.

Experiential Activities, Academic Content Review, and Formative Assessment

Current brain research supports the experiential education philosophy that educators can increase attention, motivation, and learning outcomes when they weave in opportunities to get learners away from their desks or boardroom tables and moving and interacting with their peers. Studies on attention indicate that though attention varies from person to person and can be affected by time of day and other factors, there are fewer lapses of attention during lecture and direct instruction situations when active learning methods are used and when reflective questions are posed (Bunce, Flens, & Neiles, 2010). Varying our methods for delivering and reviewing lessons helps learners create more neuro-pathways for the information to be stored and accessed later (see page 7). Getting participants away from their desks provides opportunities for learners to interact with each other socially and reflect upon the information being presented. They will likely retain it longer and integrate it better into their life and future learning.

This chapter explores active and social ways to engage learners in community building, dialogue, academic review, and reflection. These approaches can double as formative assessments, helping both educators and learners review material and check for understanding along the way. Many activities that are traditionally used as icebreakers at the beginning of a community-building program to get group members talking and connecting can be adjusted to increase attention and retention for academic settings. An effective teaching strategy is to break up a lecture after 10 to 20 minutes of direct instruction and get learners away from their seats, moving and reflecting on the material just presented. The following activities serve both purposes. What works well as a group-building-introductory activity at the beginning of the year can be re-purposed as a reflective or academic review strategy throughout a program or course.

The techniques described in this chapter work well as a second step to the object-based entry tasks discussed in chapter five. The strategies outlined in this chapter allow group members to share about their choice or entry task with one person at a time before being asked to share with the entire group therefore building comfort with reflection. These are great methods for sparking conversations in the classroom and an alternative to the more didactic question and answer sessions, as everyone is involved in reflecting, talking and sharing rather than just the hand-raisers. These social learning strategies increase engagement, develop multiple neuronal pathways to learning, and promote social-emotional skills.

Concentric Circles

Purpose/Focus: active engagement, movement, academic review, formative assessment, group discussion, differentiation, multiple pathways to learning, social-emotional learning, 21st century skills, communication, active listening, names, community building, reflection, peer interaction, commonalities/different perspectives, quick energizer, transition

Materials: No materials are needed; however, this is a great technique for sharing entry task "strong beginnings" choices (i.e., quotes, postcards, pin-back buttons found in chapter five).

One of my favorite tried-an-true methods for actively engaging a group in dialogue and reflection about a lesson or experience is Concentric Circles.

Facilitation Suggestions

- Divide the group in half, and form two circles with the participants facing each other in an inner circle and an outer circle.
- Don't be afraid to move your desks! With a small adjustment (e.g., pushing desks to the center), a space can be made for this formation. Make movement a habit in your class-

room, and students will learn to quickly arrange and rearrange the room as needed.

- Ask participants to greet each other and converse about the topic at hand. Prepare specific questions about the material for partners to discuss or initiate a less structured conversation by inviting participants to share their reactions to or key take-aways gleaned from a lecture.
- When discussing current events, differing perspectives, or a hot topic, have the inside circle represent one point of view and the outside circle another.
- Depending on the group, you can incorporate a fun partner activity such as Gotcha (page 192), thumb wrestling, or one handed, partner shoe-tying prior to the reflection or review discussion. This brings a little levity and appropriate social touch into the classroom, helping participants create social connections (another brain-friendly method of teaching). After completing the partner activity, ask participants to share their thoughts about the lecture, lesson or reading.
- After a few moments, or when the conversational energy diminishes, have a circle rotate a few places (depending on group size). Participants will form new partnerships, saying hello to those who they bypass. Invite the new partners to greet each other, and provide another question to discuss and/or cooperative activity.

When working with language arts classes, I often have the inside circle take the perspective of one character in a play or story and the outside circle another. I then ask them to answer questions "in character."

My colleague, Marci Charles, used this method to increase inter-department communication in a corporate team-building day. She had the administrative team form the inside circle and the IT team form the outside circle for a reflective discussion on company communication. I have tried it a few times in the academic settings with great results—once in faculty team-building with administrators on one side and faculty on the other and another time with high school mentors on one side and their faculty leaders on the other. This ensures various perspectives are being heard in situations where peers tend to group with peers.

- Continue the activity, alternating movement between the inside and outside circle, followed by questions (and partner activities if appropriate).
- I sometimes pair "themed" questions with some of the partner activities. For example, in a teacher staff-development workshop, I might pair "collaborative one-handed shoe tying" (a partner task) with a request for the teachers to share their experiences with co-teaching.

Academic Variation

Purpose: active engagement, differentiation, conversation starter, review, movement, formative assessment, social engagement, reflection, accessing and analyzing information, community building

Materials: 3 x 5 index cards with questions, pencils, *The Odyssey* Study Guide, *The Odyssey*

Mark Upright, a high school English teacher who attended a workshop of mine, shared how he adapted Concentric Circles for his study of *The Odyssey*. He gave the inside-circle students index cards with questions for quizzing the outside circle and marked correct and incorrect responses as an active exploration of the material and formative assessment for both teacher and students.

Facilitation Suggestions

- Divide students into two groups and form concentric circles. Give each student an index card with a question and answer on it.
- Once everyone is in place, have the students who are facing each other ask one another the questions from their index cards.
- If the question is answered correctly, students check the "correct" category on the back of the card; if incorrectly, the "incorrect" category is checked. (The purpose of this is to gather evidence as to what areas need to be re-taught.)
- Once students have asked and answered their questions, have the outside circle move one person to the right and exchange questions with their new partners. The outside circle moves until every person on the outside circle has asked their question to every person on the inside circle and vice versa.
- In order to cover more questions, if a students answers a question incorrectly, that student switches places with his/her partner. This will allow students to answer more questions because they are part of more than one circle.
- Students can trade their question card if they are standing in front of a student who has already answered the question.

Outcomes/Reflections

I liked this active review and formative assessment because students were able to learn the correct answers for themselves through discussions with their peers instead of relying on their teacher for the correct information.

Resources/References: I was first introduced to this activity as an icebreaker facilitated by my friend Hutch Hutchinson (Cain et al.). I re-purposed it over the years as a reflection and review activity with impressive results in the classroom and with team-building and training groups.

Handshake Mingle

Purpose/Focus: active engagement, group discussion, movement, community building review, reflection, differentiation, multiple pathways, communication, playful learning, and ownership, social-emotional learning, 21st century skills, names, commonalities/di tives, appropriate touch, peer interaction, transition, quick energizer

Materials: No materials are needed; however, this is another great technique for sharing entry task "strong beginnings" choices (i.e., quotes, postcards, pin-back buttons found in chapter five).

This is another playful active-partner-dialogue activity that blends popular culture with movement, social and physical connections. It is great for introductions, reviewing names, and helping a group become comfortable with each other. Like Concentric Circles, it also works as an active reflection, review, or pre-teaching activity. This is achieved by combining the handshakes with discussion questions, academic review questions, or discussion around the quotes or objects chosen during the entry task.

Despite its playful and sometimes silly nature, groups of all ages and backgrounds buy in to this activity. I believe this is because it is social, incrementally sequenced, and starts with an activity that is familiar and popular in our culture. Most importantly there is choice involved, and no one person is in the spotlight.

Facilitation Suggestions

- First, have everyone find a partner and give them a high 5. Have participants recognize this person as their high-5 partner. (Build a sense of comfort, control, and ownership by using simple, minimal touch actions throughout this activity, especially when first getting to know your group.)
- Next ask them to find new partners and give each other a fist bump. This becomes their fist-bump partner.
- Then have them find their high-5 partners, then their fist-bump partners, moving among the group to find those original partners and performing their handshake actions.
- Continue this sequence, adding in new partner activities as appropriate. Handshakes can include: high 5, high 10, low 5, fist-bump-fireworks, etc.
- Repeatedly returning to each partner is the key to successfully building rapport and connection—don't rush through this aspect of the game!

Have participant's help you come up with new handshakes or present their own. They will likely have a full repertoire! This enhances ownership and involvement.

- When using this as an introductory activity, pause in the handshake rotations and have partners share get-to-know-you questions, reflective questions that set a context for the program, or their reason for choosing their button or quote (see chapter five).
- Later on, partners can be used to form groups.
- As a closing at the end of a class or program, have participants go through the sequence and revisit their partners to reflect on their thoughts. For example, you might have high-5 partners share something from the lesson or program that was worth celebrating or a high point for them.

Handshake greetings have become part of pop culture in the US with youth and fans emulating baseball players, popularizing their "signature" greetings. There are many stories about the origin of the high-5 but the two most well-documented are about baseball players Dusty Burke and Glenn Burke of the Los Angeles Dodgers baseball team on October 2, 1977, and Wiley Brown and Derek Smith of the Louisville Cardinals basketball team during the 1978-1979 season. The low five actually showed up first in popular culture as "giving skin" in Cab Calloway's *Hepster Dictionary* in 1938.

Academic Variation

- When using this as a kinesthetic reflection or for academic review, mix up the handshakes with partner sharing about the topic at hand, points from the lecture the group just listened to, last night's reading, their lab, current events, or experiences with a class project.
- When using this in advisory or community-building programs blend reflective questions into the exercise. For example, "A high 5 is a celebratory gesture: share something that has happened at school or work recently that is worth celebrating."

Facilitation Notes

Some of my favorite handshakes are fishing partners (one is the fisher with reel/rod and the other the fish) and ankle-shake partners. I like to have fishing partners share a fishing story with each other. Despite having these favorites, I have toned down the handshakes to create more buy-in and ownership from all group members. I now start with high 5's, low 5's, fist bumps, and other simple well-known handshakes. Then I ask for ideas from the group before sharing more playful options such as fishing partners. I also blend in opportunities for choice and control. For example, if I use the ankle-shake, I will demonstrate a very simple option such as just reaching out legs and going ankle to ankle as well as the more advanced option. Use handshakes that are appropriate for your group, taking into consideration age, space, social atmosphere, and setting (but don't be afraid to experiment!). These little choices from the very beginning of a program communicate to your group that you want to empower them with a sense of choice and control over their experience. Be aware of the cultural backgrounds of the groups you work with when facilitating activities that involve touch and adjust accordingly.

The repetitive revisiting of partners is one of the most important aspects of this activity, allowing participants to create a connection with the people who are their partners and learn and reinforce names. When facilitators skip this part, it takes away from the activity and can turn a positive rapport-building activity into an intimidating icebreaker (see page 133).

Outcomes/Reflections

Group members who are reluctant to participate often buy into this activity because it starts with familiar and simple greetings like high 5 and low 5, it moves quickly, and everyone is doing the activity at the same time. Individuals can participate at their own pace or comfort level by adjusting or opting out of a handshake.

Recent research on the brain demonstrates that engaging in appropriate touch such as handshakes is important for social and emotional learning and attention (release of dopamine and increase of brain activity).

Resources/References: I learned this as a community-building activity years ago from my colleague Aimee Desrosier Cochran. A variation is described in *Tips & Tools* by Stanchfield.

Trade and Share

Purpose/Focus: active engagement, group discussion, academic review, reflection, differentiation, peer interactions, multiple pathways to learning, formative assessment, social-emotional learning, 21st century skills, communication, active listening, community building, commonalities/different perspectives, respectful behavior, names

Materials: This is a great way to use the entry task props such as buttons, postcards, or other objects. Or, simply ask group members to use a personal item from their backpack or purse that represents a goal or story, depending on the context you are setting.

This variation on what many know as the Nametag Trade icebreaker can be adapted as a playful way to engage group members in dialogue about themselves, their opinions, or goals. At a workshop where I was promoting the idea of re-purposing icebreakers as reflective approaches, my colleague Ed Caplan shared the idea of blending the "trade" activity with the entry

task objects I used when participants walked in the door. We used my set of Quotables as a "hook" or entry task that day. He suggested this adaptation of the Nametag Trade activity as a way to share our quotes and our reflections about why we chose them. Since that time I have varied the activity in many ways, using it as an introductory activity, to celebrate personal strengths, and for academic review. It is an effective way to help students learn more about each other, practice active listening, or reflect on a specific experience or academic lesson.

I find that postcards and objects often work better than the quotes in this particular activity as the objects and images are easier to remember. This activity is best when done later in group development or at least after the group spends some time together warming up with other simpler introductory activities (such as Handshake Mingle).

Facilitation Suggestions

- Ask participants to share with a partner their name and why they chose their object, image, postcard, key or button. Then have partners temporarily trade their objects with each other.
- Next ask them to visit another person and trade their new objects and the story they were just told. Do this for a total of two to three trades (more than that becomes confusing).
- After a few minutes of trading, have the whole group come back together and ask participants to introduce the person whose item they are holding. I am careful to mention that this is an experiment in active listening and not a test. This game models to participants that their fellow group members will help them out.
- Then invite the sharer to hand the object back to the person and ask them to add any additional information or clarifications to the introduction.

Personal Object Variation/Trade and Share

Lynn Sweet, a high school music teacher from Bennington, Vermont, shared her variation: "I didn't have computer keys or anything for the students to trade, so I just had them grab something from their bag or pockets to represent why music was important to them. They used phones, jewelry, a water bottle, a book, whatever they could find. It worked really well to express their ideas. They loved it, and they had some really good feedback about the activity."

Academic Content Variation/Trade and Share

I asked teachers at Mt. Anthony Union High School to explore ways they might use this activity with their students. Jessica Dubie, an English teacher, tried this version in her classroom:

Materials: 3 x 5 cards—each with a character's name on it, *The Odyssey, Oedipus the King,* pencils

Directions

- Scatter cards with the character's names on a table, and have each student choose one.
- With the allowance of using their books, ask the students to write one sentence that identifies their character on the card.
- When everyone is finished, invite students to find a partner, share their identifying statement, and guess each other's character. Then have them exchange cards.
- Students add another identifying sentence to their new cards, find a new partner, and repeat the process.
- They exchanged cards three times—many students ended up with their own cards again, which was actually good because a few of the characters were quite obscure or from a piece that had been read some time ago.

Outcomes/Reflections

Jessica reported that it was a great formative assessment. With each round, students had to expand upon the character, looking deeper into the text. The cards can be used again to play Celebrity, where students state the information in front of the class and the class has to guess the characters.

Academic Variation Creating Context and Inspiring Inquiry/Trade and Share

Jessica Hammond, a science teacher at Twin Valley Middle School, asked if we could try something to draw students into the unit she was starting on Earth's atmosphere. Her hope was to inspire thinking around how the atmosphere affects students and the life around them. We facilitated this activity on the second day of the unit after Jessica had briefly introduced the basics of the topic.

Materials: index cards, pencils, and a chosen assortment of my Pick-a-Postcards, Chiji Cards, and Miniature Metaphors that represented living creatures, plants, and natural phenomena such as snowflakes, sun, rain, etc.

Facilitation Suggestions

- Lay the objects and images out on lab tables. As students enter the room, invite them to choose an object or image that interests them.
- Hand everyone an index card. (As with other object based entry tasks, students are drawn into the lesson from the moment they walk into class.)
- Ask students to take a moment and think about their object or picture and what they know about how it interacts with the earth's atmosphere. Ask them to write down three things:
    1) How does your animal/plant/natural phenomena affect the atmosphere?

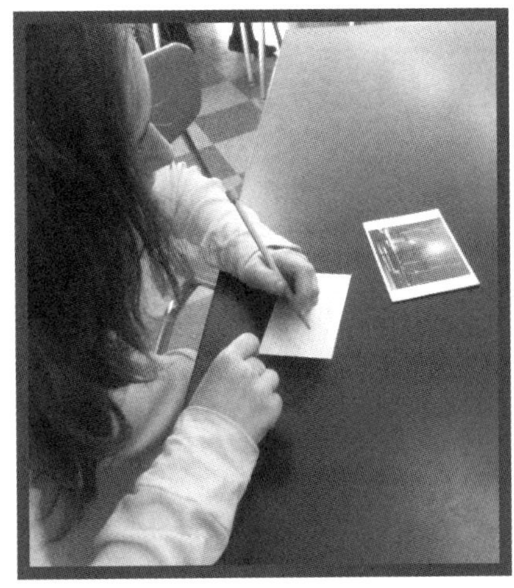

2) How is your animal/plant/natural phenomena affected by the atmosphere?
3) What question(s) do you have about how this creature or phenomena interacts with the atmosphere?

- Then go through the Trade and Share sequence. In this variation, ask students to take time during each partner object trade to add notes from their discussion with their partner to the index card. (For example, for each trade they might add an additional bit of knowledge that came up in the conversation with their partner around how their item interacts with the atmosphere. Or, after talking with their partner they might add another question about how that item and the atmosphere interact.)
- The index card is then sent with the object to the next partner.
- After two to three trades, ask students to come back the center of the room and form a circle to share the cards. Students share the item they ended up with, what knowledge was collectively shared about its interaction with the earth's atmosphere, and finally any questions that were added to the card.

Outcomes/Reflections

There were a number of positive outcomes from this activity. It became a formative assessment for Jessica and the students to understand what their base knowledge of the earth's atmosphere was at the beginning of the unit. A great list of questions was generated by students about the atmosphere and how it interacts with and affects life on earth. It seemed to spark student's interest in exploring the subject further. This activity also became another important opportunity to practice social-emotional and community-building skills such as interacting with people outside of one's friend group, and practicing pro-social communication and problem solving. Try this again at the end of the unit as a summative assessment!

The Trade and Share variations given here are just the beginning—the possibilities for creative learning are endless. For a variation that focuses on personal strengths and community building see page 170 of chapter eight.

Resources/References: Nametag Trade is a commonly used activity. Ed said he was inspired by a similar activity, That Person Over There, from *The Chiji Guidebook* by Cavert and Simpson.

Dominos and Other Simple Prop "Match Ups"

Purpose/Focus: dividing groups, active engagement, group discussion, reflection, community building, collaboration, differentiation, academic review, formative assessment, math, social-emotional learning, 21st century skills, problem-solving, peer interaction, commonalities/different perspectives, playful learning, multiple pathways, transition, names, choice, control and ownership

Materials: Dominos of any kind—double 12s and sets with interesting colors and numbering options.

Dominos are a playful way to engage participants as they enter the room and a great technique for creating partners for pair-sharing, problem-solving activities, or reflective dialogue. This partnering method stretches participants to interact with a number of different people in the group.

Facilitation Suggestions

- Hand everyone a domino as they enter the room. (I am amused by how excited students of all ages are to receive a domino—in this era of digital games, they are a novelty.)
- Ask participants to match up their dominos (dots or colors) with another person. I challenge group members to make sure EVERYONE has a partner. This means there is some problem-solving involved and interaction with a number of different people in the group.
- Have partners discuss a reflective question, a get-to-know-you question, their opinions on a current events topic, last night's homework, or their reactions to the lesson just presented.
- After a few moments of discussion ask the partners to find another pair who matches them in some way—this is where they get creative (using divisible numbers or turning the domino over). Then have the groups of four share about their previous conversations and offer another question. This naturally progresses into a whole group discussion.

Academic Variations

- This method works well to initiate reflection and group discussion around current events. For example, I used it with 8th graders in a social studies class after the earthquake and tsunami of March, 2010 in Japan. The students knew very little about what happened, so we focused on the event and its consequences that day. Students shared with each other what was known about the earthquake and tsunami in Japan and how they felt about it. After we moved into groups of four, we asked them to compare their conversations with each other and also discuss why an event happening thousands of miles away mattered to us here in Vermont. This organically led to a full class discussion as the groups started asking questions of each other and the adults in the room. Participation was active and dialogue was rich. The chance to formulate and practice sharing their thoughts, first with partners and then small groups, heightened their ability and interest in contributing to a large class discussion.
- Another example of an academic variation uses keys (in place of dominos) to engage learners in conversations about positive group norms and personal accountability to the learning community. I visited a hardware store and picked up a bunch of key blanks—two each of the

same shape/color to use as a way for students to find matching pairs for discussion. The students had just completed a classroom constitution. The keys helped encourage discussion on "keys to making the school year a success" (see page 171) and reflection on what they each thought was the key to making their classroom constitution work. This jumpstarted a productive large group discussion on accountability. On a whim at the end of class, I invited students to take the key with them as a reminder of their commitment. I noticed a few of them tying their keys to their shoelaces or attaching them to their backpacks.

Variations

There are many other game pieces and props that can be used in a similar way to the dominos.

Memory Game Card Match Up

I use a couple of memory match games when I want to quickly engage students in paired sharing about the previous day's lesson or when I introduce questions to set context for an upcoming activity. As students walk in the door, I hand them a match game card and challenge them to find their matching partner and share their thoughts with each other regarding the reflection or review question at hand.

Popsicle Stick Partners

Teacher Keli Rosso Gould uses assorted paired popsicle sticks with shapes or letters printed on them. She mixes them up in a can or jar and uses them as a way to divide partners or choose students for an activity or task. As students walk in the door, they draw their stick out of a jar and search the room for their "popsicle stick partner."

Key Blank Partners

This version is a great set up for discussing "key" points of any kind (i.e., keys to making the school year a success). I visited the hardware store and picked up a bunch of key blanks—two each of the same shape/color to use as a way for students to find matching pairs for discussion. See page 171 for "Keys to Success" group norms reflection prompt.

Playing Cards Groupings

Kristyn Harrington, a guidance counselor at Mt. Anthony Union High School, uses standard playing cards to group participants for dialogue and reflection. Just arrange the deck to have an equal number of cards from each suit and have participants group up by suits. To switch groups so they are talking with others in the class, have them then find groups that add up to a certain number or put them in straight or flush orders.

Facilitation Note

The novelty of using objects draws participants into the lesson or experience. I notice that participants are more open to being paired with someone outside their social circle. As the facilitator, you can discreetly arrange certain pairings if needed by subtly handing paired items to selected individuals. Hopefully this will inspire you to re-purpose items from your junk drawer and game closet, and pick up items at yard sales to use in this way.

Resources/References: I first started using dominos as an entry pair-sharing activity when co-facilitating workshops with Michelle Cummings. This activity was inspired by the Onimod activity used by Sam Sikes. See Sikes' *Executive Marbles and Other Team-Building Activities.* Michelle's variation is described in *A Teachable Moment* by Cain et al.

Commonalities Mingle

Purpose/Focus: active engagement, group discussion, context setting, celebrations/appreciation of others, community building, conversation starter, collaboration, reflection, communication, peer interactions, commonalities/different perspectives, formative assessment, differentiation, playful learning, energizer, organizing groups, dividing groups, choice, control and ownership

This is an effective activity for advisory groups, morning meetings, team-building sessions and a useful method to reflect on academic material in the classroom while learning about other students and exploring common perspectives. At the start of the school year, a few college faculty members posted queries on my blog looking for simple, large group, introductory activities that would help them not only establish rapport but also gauge their group and move them quickly into the content portion of their courses. This activity can facilitate the understanding of "who is in the room" and help participants make connections and start dialogue.

Facilitation Suggestions

- Invite participants to mingle (move around the room or space) and then give a prompt for them to form groups by a category or topic. A good attention-getter is useful in this activity (see page 62).
- Depending on the group's make up, age, or purpose for being together, ask participants to form groups with "get to know you" prompts such as
    - People born in the same season as you.
    - People who like the same genre of movies as you.
    - People who like similar sports/activities.
    - For some humor, one of Anthony Curtis' prompts: "People who used the same toothpaste as you this morning."

- When trying to determine who is in the room, have them form groups by role or position at school or reason for attending the training.
- This activity moves quickly and doesn't need to take more than about 10 minutes.

Variation

I often use this in combination with the entry tasks shared in chapter five—asking participants to form groups of people whose buttons are similar in some way, or whose postcards or quotes have similar themes. As they work to form groups, participants engage in reflective dialogue around the interpretation of their postcard, quote, or object. An understanding of differing perspectives and common themes emerge.

The report-out or explanation shared by each group is the most important aspect. Usually a representative from each group will share their theme or commonality with everyone. Through this process participants who haven't yet found the "right" place are welcomed into groups. This discussion process enables everyone to find out more about who is in the room, make connections around shared goals or commonalities, and potentially gets everyone talking and reflecting about topics to frame an upcoming lesson. It is always interesting to me how participants find each other, define these commonalities, and invite others into their groups. I like the reflective discussions that organically arise around these groupings.

Academic Variation

In an academic situation, have students group up by their perspectives on a character or event in history, their perspectives or reactions to a current event, their experiences with technology, or their predictions around an outcome to a problem being posed in class.

Facilitation Notes

Increase comfort and buy-in by starting with partner or small group sharing activities. This gives participants an opportunity to warm up by interacting with just one or two others before sharing with the large group. Blending in reflective questions helps participants focus on a topic from the very start of a program leading to more in-depth reflection discussions later on.

Resources/References: I learned the commonalities activity from Anthony Curtis of Adventureworks. He attributes his variation, which he calls "Incorporations," to Weinstein and Goodman (*Playfair).*

Anyone Who (Re-Purposed Version of Have You Ever?)

Purpose/Focus: active engagement, academic review, movement, formative assessment, reflection, group discussion, differentiation, multiple pathways to learning, community building, collaboration, context setting, social-emotional learning, 21st century skills, commonalities/different perspectives, playful learning, turn taking

Materials: spot markers for every player, review or reflection question cards

In re-purposing the Have You Ever activity for reflection and academic review, I discovered the Anyone Who variation works as a kinesthetic way to review academic material in the classroom, assess a group's knowledge around a subject, pre-teach, or initiate reflective discussion.

Facilitation Suggestions

- Provide a spot marker for every person minus one in a circle on the floor.
- Then place a spot of a different color in the perimeter of the circle. This becomes the "questioning spot."
- The person on the questioning spot asks a review/reflection question about the experience or topic at hand.
  - For processing or reflection, group members can pose reflective statements about an experience such as: "Anyone who tried something new today," "Anyone who stepped up as a leader," etc.
  - For academic review, prepare relevant questions on an index cards such as: "Anyone who knows the difference between a metaphor and simile," "Anyone who knows the freezing temperature of water," "Anyone who can name one thing the atmosphere does for us here on earth," "Anyone who can describe the process of photosynthesis," etc.
- Then everyone who has an answer to the question leaves his or her spot and tries to find a new one.
- Someone new ends up on the questioning spot and asks another question of the group.
- I add more opportunity for choice and control within the game by providing a buzzword such as "Bananas" for the person on the hot spot to use if they can't think of a question to ask. When this buzzword is used everyone has to move.
- There are a number of ways to share the questions and answers in the game. The person who ends up on the questioning spot could share the answer when they arrive there, or they could ask for input from the others who moved.
- I have participants design the rules of the game in a way that works best for their group.
- If I feel the group is ready for it, I will invite participants who end up on the questioning spot to share anything they would like to about the question/experience at hand. Most will eagerly share a story or vignette.

The first time I used this method was with an 8th grade social studies class studying mid-nineteenth century US history. Topics included Westward Expansion, the American Civil War, and the Reconstruction Era. The teacher and I pulled out key events, people, and facts to formulate review prompts such as: "Anyone who knows who Jefferson Davis was." "Anyone who knows the outcome of the treaty of Guadalupe Hidalgo." "Anyone who knows what the Pickwick Papers were." "Anyone who can name the confederate states." We wrote them on index cards and placed them on the questioning spot.

Many students couldn't help blurting answers out while they moved. If a person didn't have the correct answer, other students immediately jumped in to help in a friendly and supportive way. When one student moved impulsively from his spot without knowing the answer and reached the questioning spot, he froze for a moment and then asked me if he could "phone a friend." Immediately, other students pretended they had open phone lines to help out their classmate. This "phone a friend" method became something we used the rest of the year when someone needed help. It even spread to other schools and programs when I shared the story with group facilitators and teachers in trainings who then shared it with their students and participants.

Students who usually didn't raise their hands in class were actively moving and sharing answers throughout the game. The teacher gained new insights into the more introverted or previously less-involved class member's actual knowledge of the subject. It was great for the high-achieving students in the group (who sometimes monopolize class discussions) to see that other students had knowledge and insights to offer the group.

As the game progressed students started to move beyond review questions and answers, spontaneously engaging in detailed discussions around topics such as the impact of the railroad on the Civil War, and how the Westward Expansion had affected modern-day population and industry in their own state of Vermont. The best part was that students were leading the discussion!

Variations

- I have also used Anyone Who in counseling, guidance, and leadership groups by reviewing factual information or introducing a subject with prompts like the following:
  "Anyone who has a definition for empathy."
  "Anyone who can name one way to stop a fight without violence."
- Sample prompts for a peer mentoring group of high school students who are working with younger students in after school programs:
  "Anyone who can describe how they communicate their role or take their space as a student leader when the program starts."
  "Anyone who can name one important aspect of communicating directions for an activity to participants."

"Anyone who can describe a situation where they might seek support or help from an adult leader."

- Some prompts for teachers in staff training or for facilitating advisory groups:
  "Anyone who can share a success story from his or her advisory group."
  "Anyone whose teaching practice has improved since they became an advisor."
  "Anyone who was challenged by his or her advisory group."

Facilitation Note

This has become one of my favorite and most commonly-used activities to start reflective conversations in advisory or peer leadership groups or to review academic material in class. It is a great way to start a dialogue about reactions to a reading assignment or to review material right after a lecture or direct instruction. This activity can be a method for assessing knowledge of a subject within a group or as a technique to introduce a subject prior to a lesson or group experience. It is a great example of an active and engaging formative assessment. The level of involvement and discussion almost always goes beyond my original expectations.

Though this activity is great for factual review and as a prelude to a lesson, I am cautious when using it or any game to discuss feelings and experiences around sensitive issues such as bullying or substance abuse. Using a playful game to share feelings about tough issues can potentially trivialize these serious and sensitive subjects, put group members in an emotionally unsafe environment for sharing, or cause a group to cross boundaries in a way that can be inappropriate for school settings. As with any activity, carefully consider how it fits with your group's personality, development, goals and setting for best results.

Many educators report to me that they like Anyone Who because everyone is involved in the review even if they aren't speaking out or moving from space to space. Those passively participating can still be learning from the dialogue and review.

The Have You Ever/Anyone Who activities can be used multiple times with the same group. I like to use the Have You Ever? version earlier on as a get-to-know-you activity. Then later in the group's process, re-introduce the game as Anyone Who with the intention of reflection or review. It doesn't matter if a group has played the game before, it is different every time. When participants have familiarity and comfort with the structure of the game itself, they are more willing to push their comfort zone with the questions at hand.

Resources/References: This is a variation of Rohnke and Butler's Have You Ever activity, *Quicksilver,* 1995. Thank you to Patrick Torrey for inspiring a reflective version of this activity in *A Teachable Moment* (Cain et al.).

Classroom Four Square

Purpose/Focus: active engagement, academic review, formative assessment, movement, multiple pathways to learning, playful learning, social-emotional learning, executive function, self regulation, turn taking, fair play, peer interaction, quick energizer

Materials: tape, playground, tennis ball, review question cards

An innovative teacher, Anne Sulzmann, has taped a Four Square court in her middle-school classroom. The classic, playground game is popular with middle-schoolers, and it is one of the few group games they spontaneously start up during free time outside. She uses the game as an active way to engage students in review. When a person is tagged, they have a chance to answer a review question to stay in the square. The game moves quickly so everyone stays engaged. The spectators quickly move in and out of the game, and like the Anyone Who activity, all the participants are reviewing the information whether it is their turn to answer or not!

1	2
4	3

Facilitation Suggestions

- Tape a square on the floor with sides of at least 6 feet, add strips of tape to divide it into four equal squares, and then add the numbers (see diagram). If you are doing this in the classroom, desks can be rearranged accordingly—I recommend keeping it taped to the floor for ongoing use.
- One player stands in each square, the rest of the class waits in line behind square one to enter a spot.
- The object of the game is to eliminate players in higher squares so that you can advance to the highest square yourself.
- To start the game, the player in square four serves the ball by bouncing it in her square once and then hitting it, flat-handed, toward one of the other squares. The ball must bounce in another player's square and he must hit it to another player before it bounces a second time or he may hit the ball before it bounces. The receiving player hits the ball to any player in one of the other squares.
- If a player hits the ball so that it misses another player's square, or fails to hit the ball before the second bounce after it has landed in her square, she moves out of the game.

- When a player moves out, the other players move up to take his or her place, and that player moves to the end of the line.
- The academic review version moves quickly. When a person misses, instead of moving to the end of the line, he or she has a chance to answer an academic review question. If a correct answer is given, he or she gets to stay in the game.
- All the participants are taking part as they listen to the review Q & A.

Facilitation Notes

I was so impressed by the level of engagement in this active review in Anne's class, I mentioned it during a training to high school teachers. One high school teacher started using it in his science class. Students now ask to play a round to review before a test.

Pass the Knot

Purpose/Focus: academic review, reflection, group discussion, formative assessment, differentiation, multiple pathways to learning, social-emotional learning, community building, playful learning

Materials: large rope or piece of webbing with a knot tied in it and then tied in a circle

This simple engaging conversation activity adds a little playful randomness into a traditional Q & A reflection or review session.

Facilitation Suggestions

- Have your group circle-up around the rope.
- Explain that they will be gently passing the rope (so as not to burn hands)—passing the knot around the circle from student to student.
- When the facilitator says "stop," whoever has the knot closest to or in their hand shares a reflection or answer to a specific reflection or review question.

Facilitation Note

I find youth enjoy this activity as opposed to a traditional sharing circle because of activity, playfulness, and the random choices of who might answer. I always allow and encourage peers to help each other out with sharing.

Resource/Reference: I learned this from my friend, Josh Meyer. Tom Smith shares variations of this activity in his Raccoon Circles workshops, see *The Book on Raccoon Circles* by Smith and Cain.

Review and Reflection Dice

Purpose/Focus: academic review, reflection, conversation starter, formative assessment, playful learning, group discussion, vocabulary, alternative to Q&A session

Materials: Chiji Processing Dice, dry erase dice, small boxes, or recycled styrofoam

This activity uses a reflection tool created by the makers of Chiji Cards (see resources). These Styrofoam dice have questions that cover "what happened," "so what," and "now what" and then a fourth die with arrows deciding who gets to answer. I find that elementary and middle school students love to roll the dice. The trick is to roll the question die before the one that decides who is going to answer so that everyone reflects.

Resource/Reference: You can easily make these yourself for academic review and reflection. A number of educational supply companies sell "dry erase" dice, which can be written on and erased. Many educators I work with recycle small boxes or pieces of Styrofoam to make large dice. Chiji Dice are available through chiji.com and experientialtools.com

Gallery Walk

Purpose: active engagement, collaboration, reflection, content review, group discussion, goal setting, differentiation, formative assessment, conversation starter, commonalities/different perspectives, movement, vision setting, transition, community building

Materials: flip-chart paper or pieces of butcher paper, markers for everyone in the group, tape

I find the Gallery Walk activity a useful tool when I want to gather information on experiences, opinions, and questions from group members. It is a simple and effective method for getting a group moving, reflecting, and engaged in a topic while making sure everyone's voice is represented. It is a helpful technique for generating a list of group input on a subject (i.e., solutions, goals, questions) to use as a reference later in the training or for goal setting.

Facilitation Suggestions

- Tape a number of large pieces of paper on the walls or table tops, spaced apart so that participants have to move around the room.
- Label each chart with a review question, statement, or issue related to your topic.
- Give participants markers and ask them to move from paper to paper and write their responses/input.
- This can be done quietly as individuals or as partners to increase interaction and comfort with the process. Doing this in pairs generates more discussion.
- When the papers are full of comments, take the participants on a gallery walk or tour of the room, reading and discussing their comments.
- I sometimes do this touring in pairs or small groups before opening the discussion to the whole group.
- Questions for the group might include
  - What did you notice as you read the charts?
  - Are there themes or patterns that keep surfacing?
  - Was there something that surprised you?

Variations

Enjoy and learn from a real Gallery Walk, using artwork/drawings.

Resource/Reference: I learned this activity during a vision-setting meeting with Pauline Chandler of Antioch University New England. Versions of this activity can be found in a number of books, including *The Ten Minute Trainer! 150 Ways to Teach It Quick and Make It Stick* by Bowman.

Can a Lecture Be Experiential?

While in graduate school, I had an interesting conversation with Jasper Hunt, my professor at Minnesota State University, Mankato. We were filling out conference proposal forms for an experiential education conference. He commented about the "check box" on the application form requesting us to identify what portion of the presentation would be experiential and what portion would be lecture-based. Jasper shared that he felt this was counterproductive to defining a quality presentation. He argued that a good lecture CAN be experiential.

I reflect back on that comment when attending (or delivering) a lecture or lesson. Recently this topic came up in my work with both classroom teachers and corporate trainers who struggle with the need to cover a great deal of curricular content in a short time in a structured classroom or boardroom setting. These educators struggle to balance their need to cover the content and their desire to teach more experientially in order to engage learners and create meaningful and lasting lessons.

Last summer, I attended two keynote lectures at educational conferences that were very engaging and left a strong impression on me. Surprisingly, they did so in what might not usually be considered an experiential format for teaching: a giant lecture hall and a PowerPoint® presentation. I reflected on that conversation with Jasper and wondered: What was it that made those lectures engaging? Would they be considered "experiential"?

When I compared these effective presenters' actions with the principles of experiential education, I found they incorporated many tenets of experiential education and brain-based teaching strategies:

Emotional Engagement: When the speaker is knowledgeable and passionate about their subject, it comes through regardless of the format of their lesson. This kind of energy was contagious and could not help but cultivate some sense of relevancy and buy-in from audience members. They encouraged and acknowledged comments from individuals in the audience, and/or walked around the room rather than just standing at the podium. When Jasper lectured on the subject of experiential education philosophy or ethics, his interest and passion around the subject were obvious and infectious. I remember being drawn in by his enthusiasm and class time flying by.

Recently, I worked with a group of high school teachers who were trying to teach more experientially in an atmosphere with a strong

The **flipped classroom** is a model of teaching in which the typical lecture and homework elements of a course are reversed. Students engage in lecture and acquire basic content outside of class time, using pre-recorded online lectures. Class time is used for inquiry into lecture content, applying knowledge, practicing skills, and interacting with one another in dialogue and hands-on activities. The activities in this book will fit well into the strategies of an educator implementing a flipped classroom.

emphasis on test scores. I challenged them to reflect on what it was that first ignited their passion around the subject area. ***If they are in touch with that passion and find a way to communicate that, I believe they will naturally find a way to stimulate some of that same passion in their students.***

Metaphor, Relevancy, Reflection and Differentiated Methods of Instruction: The lecturers used humor, personal stories, quotes, a provocative video or cartoon, and/or story with meaning and relevance to the learners in the room. They often asked the group to predict or guess upcoming information or analyze a piece of music or video clip. Many walked around the room and made eye contact with audience members. They used humor that was relevant to the audience and visual aids (photos, cartoons) to illustrate a concept and relate it to the audience's experience.

Transitions, Hooks, New Beginnings: The educators were intentional in the way they started and ended the presentation. They began with a reflective question, interesting story, or compelling statement to engage the group, and then shared a quote or gave the group a personal challenge to end the lecture. They offered a question and answer session and a reflective piece. In these cases, they are taking advantage of the principle of the **primacy-recency effect**—the idea that people most remember the beginning of a learning experience and the end (Sousa, 2006).

Relevancy and Transference to Real Life: The speakers infused personal stories that the audience related to, both about themselves and others. They also asked audience members to reflect on a related personal experience of their own. They spoke of problems that needed to be solved or asked questions of the audience.

Social Engagement: These speakers often asked audience members to reflect on or share an experience related to the subject with someone sitting nearby or others at their table. When they posed questions, they asked audience members to talk with a partner or with people at their table about it. When they used a PowerPoint® there were VERY FEW WORDS. The PowerPoint® served only as a place marker, prompt, or visual aid to underscore a point.

This reflection brought me to the conclusion that lectures can be experiential. With careful planning, intention, consideration of the needs of their audience, and the ability to tap into their passion around a subject, an educator can indeed engage learners experientially through a traditional lecture format. For more on this subject of brain-friendly engagement see John Medina's *Brain Rules* (2008).

Review and Formative Assessment

Whether you are a classroom teacher, college professor, corporate trainer, or counselor, incorporating movement into your lessons and discussions can increase engagement and help learners better retain and synthesize information from your lessons. Activities that were originally created as icebreakers can be easily re-purposed for review and reflection. Hopefully these methods will inspire you to think of ways in which some of your own favorite activities could be re-invented as active review strategies. In chapter seven, we will explore more active review and formative assessment methods that double as community-building, problem-solving, and social skills-building activities.

The benefits of active pair sharing and group review activities follow:

- We are social beings—when learners interact with each other socially and reflect upon the information being presented through discussion, then retention and application to future situations are increased.
- These activities support sequencing, helping to build rapport and comfort with dialogue and reflection leading to more meaningful and sophisticated engagement and sharing as the group progresses.
- Mixing up our methods for delivering and reviewing lessons helps learners create more neuro-pathways for the information to be stored and accessed later.
- Getting participants away from their desks and moving can enhance a lecture or presentation by providing an opportunity for a constructive "brain break."
- The playful aspects of these activities include appropriate physical touch and integrate popular culture.
- These activities practice collaboration, promote theory of mind and important social-emotional skills such as communication, empathy, and understanding diverse perspectives.
- The quality of sharing and meaningful dialogue often increases when people are interacting with one person at a time instead of sharing in front of whole classroom.
- Everyone is talking and/or reflecting, not just the hand-raisers.
- These dual purpose activities help you make the most of your time by engaging groups in reflection and review on content while building community and practicing important social-emotional and 21st century skills.
- Use these activities at the beginning of the year to build rapport, introduce and build community—and later in the school year for review. When students are familiar with an activity, they might be more willing to take a risk and share.

Chapter 7

Making the Most of Your Time

Multiple-Purpose Activities That Teach

Snapshot: The students in this 7th grade math class don't seem to be doing math at all. It looks like some strange version of dominos, until it becomes clear, as you listen to their conversation, that the dominos represent fractions. They are talking to each other about which fraction has the higher value and how to simplify their fractions as they work together to arrange themselves in order of lowest to highest value. Watching the students trying to figure it out, the teacher is learning what the students have learned from her lessons and where she needs to focus more attention. What strikes her most is that a student who rarely raises his hand in class, spends a great deal of time in the office, and hasn't handed in much homework lately is now taking the lead in helping the other students figure it out.

Integrating Community-Building, Social-Emotional Learning, and Academics

The teachers, corporate trainers, high school advisors, and other educators with which I work are constantly trying to create a balance of learning in their classes and groups. The challenge is to stretch their time to include building a strong learning community, facilitating reflection, and promoting important life skills along with the requirements of covering academic or training curriculum content.

When I first starting working as an experiential educator in the classroom, much of my work focused on building a positive and supportive school climate and promoting social-emotional learning. It naturally made sense to integrate the topics students were studying into some of the activities and reflective conversations. As schools again shifted toward standards-based testing, I found that integrating academic content with community building time became a necessity to validate the time spent during academics. The movement toward differentiation over the past decade helped me make an argument for these approaches, and I started realizing how effective some of my favorite problem-solving, community-building and communication-skills activities could be when review, reflection, formative assessment, and exploring and reinforcing academic content is the goal.

Proponents of experiential education promote the importance of active, as well as intellectual, engagement in learning along with the value of reflection. Educational neuroscientists' studies reveal that play facilitates learning and that educators can enhance outcomes by using interactive approaches and multiple senses to create multiple pathways to learning. Proponents of differentiation emphasize the need to vary or diversify our ways of delivering content and the importance of regularly checking for understanding or formative assessment. We also understand the need to practice collaboration in order to be ready for a career in the 21st century. In this chapter, we will begin to explore specific activities and techniques to facilitate this kind of learning.

This chapter of favorite tried-and-true activities will help make the most of your time by integrating academic content with group-building, problem-solving, and social-emotional skills development. Learners can practice, review, synthesize, and discuss academic material through playful collaborative learning. Some of the activities share a content-specific example such as math or social studies. The examples are meant to inspire your thinking about how to use this type of approach in teaching your content.

Play Dough Pictionary

Purpose/Focus: active engagement, playful learning, academic review, formative assessment, differentiation, multiple pathways to learning, vocabulary, reflection, social-emotional learning, 21st century skills, communication, collaboration, creativity/imagination, executive functions, self-regulation, competition, focus, turn taking, energizer, critical/higher order thinking

Materials: Playdough or modeling clay; paper, pencil and clipboard for the leader/facilitator

Although I first started using this activity in team-building programs, I found that it is an active, multi-sensory approach to curricular content review as well as a community and group-building activity. It is a playful approach to differentiation, and can be a fun and useful formative assessment for teachers.

Facilitation Suggestions

- Divide participants into groups of 5 to 6, using the Which One? activity on page 65.

- Have each group select a “team name” (practicing consensus/decision-making) possibly using a topic area related to what learners are studying in class or reviewing in training. Inviting group members to choose team names within a reasonable structure is a playful way to give them a sense of ownership.
- Give each group a can of play dough.
- Have each group select a first sculptor.
- The sculptors from each team go to the leader/facilitator who gives them a word (for older groups, this can be written on paper; for younger non-readers it can be whispered).
- The sculptors then return to their teams to sculpt the object for their teammates who try to guess what it is before the other teams figure out what their own sculptor is making.
- In the group-building version, the facilitator uses random words for the sculptors, such as ice cream cone, whale, bike, etc.
- For curricular content review or formative assessment, use objects and concepts from the lesson (i.e., rock formations, tectonic plate theory, geometry, parts of a cell, geography).
- The group that guesses correctly first wins that round. I have team members raise their hands when their team guesses the word as a visual cue for me as the leader to “judge” the winners during this fast paced game; however, I don’t keep score (see page 66 on healthy competition).
- Give every group an opportunity to show off their sculpture and receive appreciation from their classmates/group members.
- A new sculptor from each group is chosen, and the game continues until each player who wants to sculpt has a turn.
- This activity has choice and control built in as some participants might not choose to sculpt. Guessing is just as important. For younger groups, turn-taking is often the focus of this activity.
- I regularly have individuals switch teams throughout the game (for example every other turn I’ll ask “anyone who has blue on”, or “anyone with a spring birthday” to go to the team next to them clockwise). This increases the cooperative aspects of the activity and maximizes the movement and social interaction.

Outcomes/Reflections

This activity provides an opportunity for participants to get comfortable working with others in groups. It also sparks reflective conversations about creativity and its importance in solving everyday problems from math or science, to a conflict with a peer. I often use it to explore the concept that we all have creativity and can use it everyday, something that is not always acknowledged by learners.

With young children, the focus of the activity can be on executive function and social-emotional skills development such as taking turns and rules of play. With adult and adolescent groups, it can be a way to build rapport and initiate review and reflective discussions.

Encourage students to practice showing appreciation for other's work by applauding a winning team's sculpture, losing gracefully, and developing an appreciation for the process of play over winning. As with most competitive games, I never keep score and most groups don't even notice.

Practice and Review: Teresa Chirichella asked me to visit her 9th grade Earth Science class to facilitate group-building activities focused on social-emotional learning. I knew she was feeling pressure about meeting her science content goals as well, so we combined this with a curricular content review.

The group was studying rock formations. I asked for student volunteers who felt like they had a good knowledge of the rock formations they had been studying. Three boys readily volunteered to come up with the terms for their classmates and run the game. Teresa and I watched on the sidelines as they facilitated the game for their peers who were actively engaged throughout. The interesting thing was, the student volunteers were not the usual hand-raisers in class. Their teacher was pleasantly surprised at their level of knowledge. As she observed, the game became a formative assessment to gauge all of the students' readiness for the upcoming quiz.

Resources/References: I adapted this from an activity in *Quicksilver* by Rohnke and Butler who credited Ann Driscoll of the University of New Hampshire Browne Center. This variation appears in *Tips & Tools* by Stanchfield.

Charades Race

Purpose/Focus: active engagement, playful learning, academic review, formative assessment, differentiation, movement, multiple pathways to learning, vocabulary, reflection, social-emotional learning, 21st century skills, communication, collaboration, community building, creative thinking, executive functions, turn taking, self-regulation, focus, fair play, energizer, innovation

Materials: Index cards or paper and pen for facilitator, a clipboard can be helpful. Provide space for teams to spread out around the room.

Facilitation Suggestions

- Put together a list of at least as many concepts or words as there are members on each team.

- Divide participants into groups of 5 or 6, using the Which One? activity on page 65.
- Some team members might have to act twice, and some might choose not to act and send another in their place (this built-in choice helps make the game work, though most people choose to act).

- Give groups time to strategize. For example, they can share charades signs for better communication.
- Have the first actors get their word from the facilitator. When teammates guess the first word, the next actor goes to get the next word.
- The first team to get through the whole list wins.
- In order to keep track of which word each team is currently trying to guess, make sure to have the new actors tell you the word their team guessed last.
- It can be fun to have teams show how they guessed/acted out some of the more difficult concepts. Most groups spontaneously start asking and sharing with the other teams about how they communicated more abstract ideas. This can lead to meaningful reflective discussions around the content, communication, and creativity.

Success Moment: A middle school teacher asked me for an interactive way to initiate a class discussion about a book students were reading called *The Giver* (Lowry, 1993). Charades Race came to my mind. Because I was unfamiliar with the book, I asked students to each write a person, place or thing from the book on a paper and pass it to the front. The teacher and I chose the concepts while students strategized with their teams. I was impressed with their buy-in and ability to communicate abstract concepts from the book.

The best part was the reflective discussion that the students initiated when the game ended. They asked other teams how they acted out and guessed the more difficult concepts. This led to an in-depth discussion around their reactions to the book. The teacher reported back to me that students who had not been keeping up with their reading prior to our game were suddenly showing an interest in the book.

Outcomes/Reflections

Charades Race works well for characters in a book, events, or theoretical concepts.

Enhance learning and ownership by giving participants the opportunity to lead the game. Ask if there are participants in the group who feel like they have enough knowledge of the concepts to come up with a review list for the game. This works best played in a second round when they understand the game. Teachers are often surprised by who volunteers and find out through the game who has mastery of the content.

Resources/References: This activity was inspired by a charades relay game, FEACH, I learned from my colleague Karl Rohnke who credits his co-author of *Quicksilver* Steve Butler for inventing the "Fast Foods, Electrical Appliances, and Comic Book Heroes" pantomime game.

Bag of Nouns

Purpose/Focus: active engagement, academic review, community building, collaboration, descriptive language, communication, active listening, differentiation, formative assessment, multiple pathways to learning, vocabulary, playful learning, social-emotional learning, 21st century skills, creativity/imagination, executive function, focus, turn taking, self-regulation, fair play

Materials: scrap paper, pencils, timer, and a bag, bowl or other container

Similar to Play Dough Pictionary and Charades Race, Bag of Nouns is a community-building activity that doubles as an active training or academic content review and formative assessment. The game involves three rounds and can take a fair amount of time.

Facilitation Suggestions

- Write nouns on pieces of scrap paper and place them in the container. The leader/facilitator can write up a series of nouns having to do with an academic subject or training content. Or simply ask group members to each write a noun on a piece of paper and place it in the bag. Prepare at least three nouns per person so there are plenty of words from which to choose.

- Divide the group into teams (depending on the size of the group, 2 to 4 teams work well).
- The game includes three, timed rounds.
- In round one, the team who is up first chooses a player to draw a noun from the bag. The player then describes the noun to their teammates without saying the word on the paper. As soon as their teammates guess the word, the player draws another from the container.

- This goes on until the timer reaches one minute. When the minute is up, the team adds up all of the nouns they were able to guess and receives a point for each one.
- The next team repeats this with the remaining words in the container.
- Depending on time, play one round per team or play until each player on each team has had a turn describing the nouns to teammates.
- Throw all the nouns back into the container.
- In round two of the game, players proceed in the same way; however, they are limited to one word descriptions. Teammates should try to remember the nouns from the previous round to help them guess.
- In round three, players must make their teammates guess the noun without using any words at all, but by acting out the word instead. At the end of the three rounds, the team with the most points wins.

Resources/References: A number of teachers who have attended my workshops share this as one of their favorites. There are variations posted on word game and party game websites.

Zoom Problem-Solving and Communication Challenge

Purpose/Focus: social-emotional learning, 21st century skills, communication, community building, collaboration, problem-solving, creativity/imagination, executive function, active listening, patience, managing frustration, focus, sequencing, leadership/followership, higher order/critical thinking, writing prompt, differentiation, multiple pathways to learning, descriptive language, big picture perspective, different perspectives, peer interactions, persistence

Materials: The book *Zoom*, by Istvan Banyai, cut apart and laminated

This is my take on the popular team-building activity, *Zoom*, with variations that connect it with academics and curricular content review. *Zoom* is a picture book illustrated by artist Istvan Banyai (1995) with a series of single-sided illustrations that zoom out from a rooster's comb to a view of the planet Earth from space. This tool is used by many in the team-building and adventure education field. The book is taken apart, and the pages are used for this group problem-solving activity. *Zoom* inspired the curriculum content variation the follows.

Facilitation Suggestions

There are probably as many variations as there are facilitators; here is my favorite:

- Hand a page of the *Zoom* book to each group member.
- Explain that the page cannot be shown to anyone else, but that group members can talk about and describe their page to others.
- Challenge the group members to put themselves in the order of their pictures without showing them.

Facilitation Notes

This activity is an effective tool for sparking group reflection about how they communicate with each other, the varying perspectives and styles in their group, how they deal with frustration, and what roles they take in the group and leadership.

When used with school faculty, it can lead to interesting reflections on how teachers often group by grade levels or subject areas instead of looking at the bigger picture of school-wide goals.

In the business world, the reflective discussion can center on the roles participants took in the game and how it relates to their roles in the workplace as well as the lines of communication that exist or need to be improved.

With students, the lessons often center on listening to others, making a plan, designating a leader, as well as dealing with frustration and perspective taking.

Regardless of the group and setting, this activity brings to the forefront behaviors and dynamics that challenge the group and lead to rich reflective discussions. Keep this in mind when you try it, and set aside enough time for the reflective process necessary to take advantage of the teachable moments that arise.

Variation for Reviewing Curricular Content

Use *Zoom* in language arts classes to explore theme and as a writing prompt. One teacher I worked with had students each take a page from *Zoom* and write a short paragraph about the picture. Students then put themselves in the sequence of the written story and afterwards compared and contrasted the pictures from the book with the written story.

Resources/References: *Zoom* by Istvan Banyai (1995). I first learned about using the *Zoom* book as a team-building activity from colleagues Michelle Cummings and Chris Cavert when we co-led a ACCT conference workshop in 2002. The Wilderdom website has a nice write-up of this activity with variations at www.wilderdom.com. The Variation for Reviewing Curricular Content was inspired by my ongoing work with students at Twin Valley Middle School.

Layers of the Atmosphere and Other Sequence Challenges

Purpose/Focus: active engagement, academic review, formative assessment, differentiation, multiple pathways to learning, sequencing, reflection, social-emotional learning, 21st century skills, communication, differing perspectives, executive function, active listening, focus, patience, managing frustration, big picture perspective, problem-solving, leadership/followership, collaboration, community building, critical/higher order thinking, peer interactions
Science: atmosphere, miosis and mitosis, life cycles, scientific method
Math: order of operations, sequencing
History: timelines
Language Arts: perspective taking, writing prompts
Health and Guidance: decision making steps

Materials: images sequenced to fit the curricular topic at hand

I used the Zoom problem-solving challenge with a 7th grade social studies class to focus on community/team building and social-emotional learning goals. The success of the activity inspired me to look to other content pieces that are in a sequence (i.e., meiosis and mitosis, historical timelines, lab procedures). So when my colleague, Jessica Hammond, was teaching the layers of the atmosphere to her 7th grade science class, I put together a *Zoom*-like sequence challenge. The goal was to help reinforce and review the students' knowledge about the science content as well as the communication and problem-solving skills we were practicing.

Facilitation Suggestions

- Using one of Jessica's review worksheets that listed the layers of the atmosphere, I cut it into pieces and handed each student one with the layer of the atmosphere or something found in it.
- I asked the students to line up in order of the layers—no need to hide pictures in this version.

Outcomes/Reflections

This activity led to some great communication and problem-solving along with review, exploration, and discussion around the science content. It was interesting to see how the students applied what they had learned from problem-solving and communication in social studies to this science challenge. After completing the original Zoom challenge, they engaged in a meaningful group reflection on the benefits of establishing a group leader, taking the time to sit down and listen to each other, collecting all of the information, and trying to see the big picture. When confronted with the Layers of the Atmosphere challenge in science class, they came to an impasse as they were solving the problem. One student reminded the others of how they had handled the Zoom activity earlier in the year: "We solved it when we sat down and checked and listened to everyone else share about the piece they had so we could put it together. I think we need to do that again today." Yes, they did remember!

After that day's success, I printed out pictures of cumulous clouds, satellites, airplanes, birds, and the layers of the atmosphere, laminating them for Jessica to use in the future. I have since shared this sequencing, line-up idea with a number of educators who have used it to reinforce ideas that involve a sequence or timeline:

- Historical timelines
- Biological life cycles (e.g., frogs, butterflies)
- Parts of speech
- Alphabet and numbers (early elementary)
- Scientific method, lab procedures
- Notes/values in music
- Atoms and molecules
- Digestive system, circulation
- Procedures such as tying a knot
- Decision-making steps for making healthy choices, alcohol use, etc.
- A timeline of program successes (as a reflection on the school year)

Jean Zinn, a high school English teacher, found an old illustrated version of *Huck Finn*, copied the pictures, and facilitated a line up activity with her students to review and discuss the book. Colleen Harshburger at West Virginia University has adapted this idea to a student life curriculum involving making healthy decisions about alcohol use. In staff training situations, this activity can be used to review or start discussions about organizational procedures and protocols. It can be used as a timeline of program successes.

Resources/References: This academic review version was inspired by my ongoing work with students at Twin Valley Middle School.

Dominos Fraction Line Up

Purpose: active engagement, academic review, formative assessment, math skills, fractions, simplifying fractions, decimals, number line, playful learning, collaborative learning, differentiation, multiple pathways to learning, social-emotional learning, 21st century skills, communication, group problem-solving, leadership/followership, active listening, big picture perspective, community building, peer interaction

Materials: Dominos—double 12s make this more interesting and challenging. Choose your sets by the ages and abilities of the groups with which you are working.

I use line-up activities as fun filler or quick team-builder activities during transition times. When there are 5 minutes left in class, ask students to "Line up at the door in order of shoe size without talking," or "See how fast you can line up by your birthdays," etc.

When Yolanda D'Allesio asked me to help her 7th grade math students reflect on the frustrations they were experiencing around problem-solving, I used domino pairs as an entry-task, partner-dialogue activity. When Yolanda had to leave the room unexpectedly, the paraprofessional, Ginny Cunningham, and I improvised. She asked if we could do something with the dominos and fractions. This inspired me to ask the students to line up by their "domino fraction." I was amazed at how well it worked. Once again the non-hand raisers took the lead, helping classmates organize and simplify fractions. This was a useful formative assessment tool, revealing who really understood and could articulate their understanding about reducing and converting fractions. At the same time, students were problem-solving, collaborating, using their creativity, and practicing communication skills.

Facilitation Suggestions

- Hand each participant a domino.
- Ask them to line up by their domino fraction's value, from smallest to largest.
- I don't tell students where to begin and end—the less said by me, the more creativity they will use.
- Take this one step further, and have students convert their domino fractions into decimals. This can led to discussion around percentages.

Dominos Equations Variation

I sometimes challenge elementary students to form one whole group matching their dominos, or form simple math equations (using the appropriate dominos). This is not only an active way to partner and practice math, but it also introduces a game to the classroom that might be new to many students. Good old-fashioned dominos could be novel to a group of young students who mostly play computer games.

Telegraph and Probability

Purpose/Focus: math skills, practice and explore probability, active engagement, academic review, differentiation, playful learning, social-emotional learning, 21st century skills, communication, trust, collaboration, self-regulation, executive functioning, appropriate touch, peer interaction, multiple pathways to learning, fair play, gossip and rumors, focus, community building

Materials: a quarter, two lines of chairs (or space on the floor for a group to form two lines), a fun prop like a rubber chicken or stuffed animal, and a chart or dry erase board

Telegraph is an excellent example of an activity that can be used to focus on academic content AND important social-emotional skills and community building. It is one of my favorite ways to engage groups in a discussion about communication, thinking before acting or making a comment, and reflecting upon how one's actions affect others. It is also a playful and engaging way to explore and reinforce the concept of probability in math classes.

The full description of the activity for the purpose of social-emotional learning, communication and community building is in chapter eight. I purposely time this activity to fall during my 7th grade students' probability unit. We use it to practice the social-emotional and community skills mentioned above and reinforce what they are learning about probability.

Facilitation Suggestions

- See the Telegraph activity on page 152 for Facilitation Suggestions.
- When using the probability lesson variation of this activity, start with a short explanation of probability and predictions.
- Have students predict how many tosses will be heads and how many tails.
- When the game ends, tally and discuss probability.

Outcomes/Reflections

As an extension to the math lesson, have students research coin flip studies for homework. I recently came across an interesting story on NPR about studies by Stanford University mathematician and professor Persi Diaconis showing a slight bias toward heads and data showing that whatever side starts facing up wins 51% to 49%. The 2004 NPR radio story (Kestenbaum, 2004) and Diaconis' related article (Diaconis, Holmes, & Montgomery, 2007) can spark inquiry in your classroom around the field of mathematics and statistics. Along with exploring probability and predictions, this activity naturally lends itself to conversations around communication and miscommunication, gossiping, passing on messages without getting the right information, etc.

Resources/References: This is an adapted version of the Telegraph found in Jackson's *Activities That Teach* book.

River Crossing Math/Insightful Problem-Solving Challenge

Purpose/Focus: collaboration, math skills, logic, 21st century skills, social-emotional learning, group problem solving, critical/higher order thinking, communication, experimentation, patience/managing frustration, executive function, big picture perspective, persistence, multiple pathways to learning

Materials: scrap paper and simple props that can be used as manipulatives (colored index cards, game pieces, dominos, etc.)

The storyline that sets up this challenge involves creatures or people who are working together to get to a destination, but don't completely trust each other. One group is especially wary of the other and has agreed among themselves to never let their group be outnumbered by the other. They need to cross a river and find they have a boat that only holds up to two individuals at a time (it must be rowed by at least one person). Everyone must cross the river using the same boat, but group A must never be outnumbered by group B.

Because *The Hobbit* (Tolkien, 1937) is a popular classic, I use Elves and Dwarves as the characters for this storyline. The Elves and Dwarves are wary of each other and Dwarves are especially nervous about elfin magic. So it is easy to imagine the group of Dwarves traveling to Smaug's lair to retrieve the ring in the Hobbit, making an agreement among themselves never to be outnumbered by the Elves. Regardless of the storyline I choose, participants can't resist buying into this challenge. I use it in team-building programs, in the therapeutic setting, to help students prepare for group projects, and to practice creative problem-solving and math skills. This activity works with any even-numbered group; groups of six work well (three Elves and three Dwarves).

- A possible story line follows:
  The Elves and Dwarves start their journey to Smaug's lair to retrieve the magic stone. Because of the Dwarves' wariness of elfin magic they have made a pact to never let themselves be outnumbered by the Elves at any time. The group reaches a raging river full of Dwarf- and Elf-eating eels. Luckily there is a rowboat; however, it can only hold up to two Dwarves or Elves at a time. It must be rowed by at least one creature. Everyone must cross the river using the same boat, but the Elves can never outnumber the Dwarves.
- Provide props (i.e., dominos, index cards, games pieces) for participants to use to solve this problem.
- Larger groups can be divided into smaller groups that "compete" to solve the problem. If you use competition, somehow weave in a larger group goal (i.e., everyone is trying to save the town from the wrath of Smaug so both groups must get to his lair).

The Key

- The crux of the problem is that sometimes you have to go back to move forward. Once the group is at the point of three on each side of the river they are halfway there.
- The group is close to success when the team that doesn't mind being outnumbered (in this case the Elves) is on the opposite of the river with the boat. To finish the problem, they send one of their own by himself back with the boat. The Elf then stays on that shore and sends the two Dwarves back with the boat leaving one Elf and one Dwarf on one side and two Elves and two Dwarves on the opposite side with the boat. The one Elf and one Dwarf come back, and two Dwarves go back across. The Elf goes back by himself grabs another Elf and then goes back again to get the last Elf.

Outcomes/Reflections

I am always impressed by the level of engagement and buy-in with this activity. In a high school adventure education class, students worked on it over two class periods and stayed engaged. When they didn't solve it on the first day, I told them we would come back to it the next. They came to class the second day with diagrams and notes. Some reported that they had called each other to continue to problem-solve—others had gone to their math teacher to ask for input. This led to some meaningful discussion about how useful it can be to take time away from a problem to let it "incubate" or to look for helpful resources.

Use this activity in math class to reinforce problem-solving or with groups to explore creative thinking or the idea that sometimes you have to take a step back to order to move forward with a problem.

In a social studies class, this activity using the storyline from *The Hobbit,* sparked conversation around working with people who are different than you and building trust by spending time with them or facing a challenge together.

Resources/References: This is an adaptation of a brain teaser/folk tale that has been around since at least medieval times (9th century) and shows up in the folklore of many countries in Europe and Africa under various names/storylines.

Create a Colony Activity

Purpose/Focus: differentiation, formative assessment, application of content, multiple pathways to learning, collaboration, social-emotional learning, 21st century skills, community building, group norms, communication, planning, decision-making, vocabulary, peer interaction, creativity/imagination, innovation, citizenship

Materials: sheets of butcher paper, markers, crayons, space for small groups to spread out

This activity works nicely as a group-building activity to practice collaboration, group decision-making, conflict resolution, creativity and to initiate group norms discussions. It takes quite a bit of time (approximately 1 to 2 hours). I like to spread it over at least two class periods including time for reflection, discussion and a presentation of their colonies.

Facilitation Suggestions

- Divide the group into teams of 4 or 5 participants and give each group big sheets of butcher paper and markers.
- Share a story line:

  You and your team members are settling on this island for the long term, essentially establishing your own country and building a society. As a group, you now have a chance to create your ideal "island society." The objective is to set up the rules and government by which you will abide.
- Give the groups time to create their ideal country:
  - Name the island.
  - Design a flag to represent the qualities of your new society.
  - For what is your island known?
  - What is your leadership structure? Is it a democracy? A monarchy? A dictatorship?
  - What are your laws, rules, and/or regulations, and how are you going to enforce them?
  - Create a national anthem.
  - What is your national costume?
  - What are your national holidays and why?
- When the groups finish, have them present their island and share their artwork, anthems, rules and laws, etc.

Outcomes/Reflections

The island creations often involve whimsy, playfulness, and creativity. The collaborative, creative process, communication and decision-making involved is where the power lies in this activity. It is important to check in along the way in order to take advantage of the teachable moments that arise in the small group projects. This naturally leads to reflection about the challenges of establishing group rules and norms, and it can be a nice segue into a discussion about classroom norms or team/school rules.

Create a Colony Academic Content Variation

Using the Create a Colony Activity, I pull right from the curricular content of the class with which I am working, asking students to essentially role-play some of what the colonist had to do. Most recently, I worked with 6th graders studying the colonization of the United States.

Facilitation Suggestions

- Have small groups of students collaborate on the following objectives:
  - Name your colony.
  - Draw a map of your colony, and decide who is going to live where.
  - Identify your natural resources. How you are going to survive?
  - Identify how you are going to divide labor—who is going to do what job in order for the colony to survive.
  - Write laws for your colony.
  - Design a pact or formal agreement.
- When the small groups are finished, have them share with the larger group. Reflect on this experience and its relationship to the first colonists they are studying.

Resources/References: This is an adaptation of an activity offered by recreational therapist Alanna Jones, author of *104 Activities That Build: Self-Esteem, Teamwork, Communication, Anger Management, Self-Discovery, and Coping Skills* formerly known as *The Wrecking Ball.* The students at Twin Valley Middle School in Whitingham, VT inspired the academic content variation.

Colleen's Character and Metaphor Lesson

Purpose/Focus: active engagement, differentiation, language arts, reading comprehension, metaphor, character exploration, discussion/writing prompt, reflection, descriptive language, empathy, formative assessment, application of content, multiple pathways to learning, vocabulary, social-emotional learning, 21st century skills, creativity/imagination, communication, emotional connection to learning

Materials: Simile and Metaphor Review Sheets (see Resources/References at the end of this activity), Miniature Metaphors (or a similar collection of objects), drawing paper, pencils, markers/crayons

When working with English and language arts teachers and students, I often have students simply pick a charm or postcard and write about how that object or image represents a metaphor or simile. In chapter eight, I describe the Postcard Strength activity as an example of combining figurative language with social-emotional learning and community building (see page 169).

After attending one of my summer courses, Colleen Barrett, a teacher at Crossett Brook Middle School in Vermont, decided to use my Miniature Metaphors kit to help students better understand similes and metaphors in the context of an autobiography her 6th grade students were reading, *Small Steps: The Year I Got Polio* by Peg Kehret. The book recounts Peg's bout with polio at age 12 in 1949. Colleen chose this book because of the rich descriptive language that tells of Peg's life and death struggle and experience of paralysis and isolation. Colleen's goal was to use the metaphorical objects to help students better understand similes and metaphors in the context of the novel.

Facilitation Suggestions

- **Day One:** Review the concepts of metaphor and simile. Then ask students to work alone or in pairs to locate examples of similes and metaphors on given pages in the novel. Have students explain what was being compared and what the author was trying to communicate in the context of the story. (Colleen spent the first day of this lesson using her "Figurative Language Warm Up" worksheets.)
- **Day Two:** Greet students as they enter, using the Miniature Metaphors as an entry task. (Colleen used the Miniature Metaphors as a "hook" for students entering class on the second day. She wrote simple instructions on the board: "Good Morning! Please go to Mrs. Barrett's table and choose an object that you think best describes Peg in the opening chapters of the novel. Be prepared to share your reason for choosing the object with your table group. Give specific examples from the novel to support your thinking.")
    - Small group sharing and discussion: Have participants share with their small group about the object they chose and why they see it as a good representation of the character in the opening chapters of the novel. (Colleen allowed 10 to 15 minutes for sharing as she circulated and listened to the discussions. She was clear about the adults in the room taking an observer role so that this could be a student-directed discussion, not a teacher-led literature group.)

- Large group discussion: Open the discussion to the whole group. (Colleen has the students use specific language so they hear what metaphors sound like: "Peg is a _______________ in the beginning of the novel because...." She reminds them that there is no "like" or "as" (clues for a simile) in this exercise. They should speak metaphorically to the group.)
- Metaphor drawing: After students have finished sharing, hand out large pieces of drawing paper and ask them to draw enlarged versions of their objects on the top half of the paper, and use the bottom half to put into writing what they shared in their small group.
- Present drawings: Display finished products in the classroom and hallway, allowing students to see what classmates chose as their metaphorical object and how they thought it related to Peg.

Outcomes/Reflections

Colleen reported that several students who don't usually raise their hands in class offered to share their reasons for choosing the metaphor. She pointed out that each student attached a different metaphorical object to the character, so that students understand there was no "right answer." She purposefully connected this activity to the discussions about the author's use of simile and metaphor and how the students were speaking and writing metaphorically as the author did. She encouraged them to be "on the lookout" for similes and metaphors. In this way, students were continually revisiting and reinforcing the lesson.

Resources/References: "Figurative Language Warm Up" worksheets published by Mark Twain Media Inc. Publishers of EBooks with resources for teachers available from carsondellosa.com and other online sources. Miniature Metaphors are available from experientialtools.com.

Food Web Tag

Purpose/Focus: active engagement, movement, science, academic review, application of content, multiple pathways to learning, differentiation, formative assessment, community building, social-emotional learning, 21st century skills, fair play, self-regulation, appropriate touch, big picture perspective, playful learning, high physical/field/gymnasium game, peer interaction, vocabulary

Materials: photographs or artwork of the animal and plant species that are part of the food web being studied (laminate for repeated use or use tag holders from office supply stores)

This activity can be used to explore the concept of the food web and initiate discussion on ecosystems and the impact of a change in the food web and various species.

Facilitation Suggestions

- Give each student one of the pictures.
- Have students explore information about this animal or plant and the part it plays in the food web (i.e., What does this animal eat? Who are its predators? What animals eat this plant? Where does it grow?) This might be individual or group research. Have them share with others in the class (see partner reflection/discussion activities, pages 86-93, 95-96).

- Circle the group up and have each student share basic information about his or her plant or animal's part of the food web.
- Find an open field or gym space to begin the tag game.
- Explain that animals chase their prey and tag them. Once a prey or plant is tagged, they have to latch onto the predator and become part of that animal (a tagging team).
- To begin the game, the teacher facilitator declares: "Go Eat!" and the play begins.
- After 5 minutes, freeze the game and report out what animals were "caught."
- Have students switch roles and play another round.

Outcomes/Reflections
This activity can initiate discussion on the interdependence of plants and animals and what happens when a food source runs out.

Resources/References: There are many variations of this activity. Resources for various types of ecosystems can be found online. One useful online resource that includes a template of various animal pictures is found on exploringnature.org.

Active Coordinate Plane

Purpose/Focus: active engagement, math skills, academic review, application of content, differentiation, formative assessment, multiple pathways to learning, collaboration, problem-solving

Materials: index cards, tape, space to move around, paper, and writing utensils

The math classroom floor becomes a coordinate plane (x, y) or Cartesian plane (made of tape) in this interactive representation of a math worksheet. Students practice mapping/graphing integers to create shapes. This is a great way to review and assess after direct instruction or graph paper exercises, getting students out of their seats, moving, talking, and collaborating. I use this with older elementary through pre-algebra classes. The objective is for students to demonstrate how to identify and graph points correctly while learning and practicing math vocabulary (i.e., x-axis, y-axis, x and y coordinates, points, ordered pair, quadrants, origin).

Facilitation Suggestions

- Ask students to work together to build a giant coordinate plane on the classroom floor using tape.
- Have partners practice and assess their readiness for using the coordinate plane by giving them ordered pairs on index cards and having them stand at their graph points.
- Invite other class members to see if the location is correct.
- Next, give the students a set of ordered pairs that form a picture when they are connected. Students will test and demonstrate their knowledge by the outcomes of their pictures.
- The social-emotional learning piece comes from collaboration between teams or partners and competition between the groups.

Variations

Ask students to draw their own pictures and have other groups of students transfer them to the classroom floor coordinate plane.

Educators can use variations of the game "Battleship" to practice coordinate plane graphing with partners using graph paper. Try making this a whole class activity using the classroom grid.

Resources/References: When Chris Brown was a new math teacher at Hinsdale High School, he came up with this activity in a brainstorming session during a differentiated instruction course I was teaching. We tried it with his students who had come to the course with negative associations about math. He shared that as the unit progressed, students demonstrated a deeper understanding of the concepts.

Collaborative List Making Review

Purpose/Focus: active engagement, academic review, collaboration, playful learning, formative assessment, multiple pathways, differentiation, active listening, vocabulary, social-emotional learning, 21st century skills, turn taking, fair play, self-regulation, focus

Material: paper and pens or pencils for each group

This variation of Scattergories® and other list-making games is a great way to introduce a subject, involve learners in reviewing material through play, or conduct a formative assessment.

Facilitation Ideas

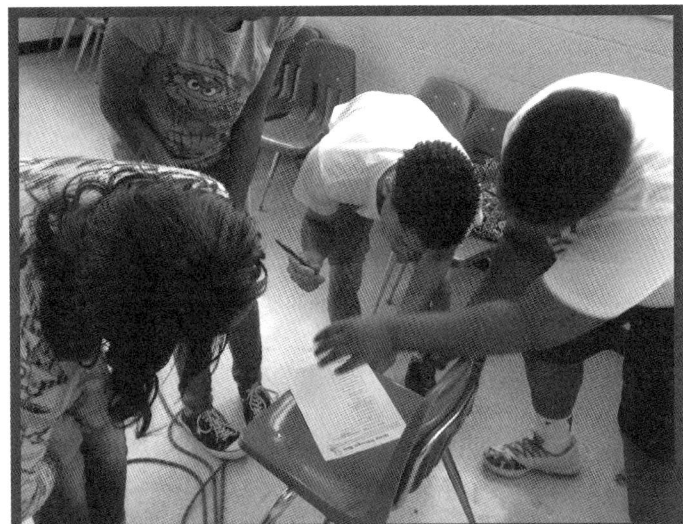

- This game is very simple. Each team works together to create a list of words or concepts under a theme, topic, or concept.
  - In a community building game, the groups might compete to see which group can come up with the longest list of celebrities whose names begin with B.
  - In language arts class, students can compete to come up with the longest list of synonyms for the word "pretty" to initiate discussions on connotation and denotation.
- As each team reports out, any duplicate words between teams are taken off the lists for the final tally.
- The winning team is the group with the longest list of words.

Variations

There are a few variations of this activity. Teammates can simply blurt out answers and have a scribe put them on paper. Or participants can quietly pass a paper and add words one at a time. It can be interesting to compare both methods to initiate a discussion on collaboration versus individual work.

Resources/References: Tom Jackson offers a community variation of this activity in his book, *Activities That Teach,* 1995.

Benefits of Using Collaborative Problem-Solving Activities for Academic Content Review

- Activities focus on academic content while at the same time helping students learn and practice social and emotional skills and 21st century skills.
- Multiple pathways to learning are being used.
- There is opportunity for interdisciplinary collaboration through the experiential activities applied to the curricular content areas.
- Students gain new insights into the curricular content by participating in academic lessons in an engaging, "brain-friendly" way.
- The non-hand raisers are engaged! There is increased involvement and buy-in from students that weren't previously active in classroom discussions around curricular content.
- All students (special education) can be included in these classroom sessions regardless of academic level.
- Educators and students get to know each other as individuals and recognize the strengths of their group members and the resources of their learning team.
- Formative assessment becomes part of the lesson itself. Both the teacher and the student increase their understanding of individual knowledge readiness or progress in regards to the academic material being explored.
- Students can take responsibility for (and even a leadership role in) their review and application.
- Learners build positive associations with learning, school and academic review.
- Students learn about the unique perspectives of their peers.
- These activities can be used at the beginning of the year to build community—and later in the year for review.

These active techniques to engage learners can be incorporated in a variety of ways throughout a course or program to differentiate instruction, engage learners in academic content, and cultivate important social-emotional skills. These ideas can be adapted to fit a variety of content areas and populations. My hope is that this chapter has inspired your thinking about how to use activities to reach multiple goals. If you are looking to expand your team-building toolkit further, chapter eight offers more strategies for building a productive learning community and developing positive group norms and social-emotional skills.

Chapter 8

Building a Strong Foundation for Learning and Life

Collaborative Games and Activities to Promote Social-Emotional Learning, Community Building, and 21st Century Skills

Snapshot: A high school teacher tried the Tin Can Pass activity with her honors 11th grade class as a way to prepare them for upcoming group projects. It was an amazing success! At first, she was nervous about trying it because she knew that many of the students in this group did not like working together; however, this activity demonstrated to the students that they COULD work together! They made it all the way to the third step of passing the can around with an object in it, with eyes closed. It did take them a number of tries and some moments of struggle. When she asked them to journal about their experience (pre and post), it went surprisingly well. The students were able to articulate the lessons learned—helping one another, being open to other's perspectives, starting again if it doesn't work out the first time, and showing empathy to the other participants (something this group of students really needed to practice). She was so excited to have found the perfect way to start off the unit; she will be able to refer back to this activity as they progress through their projects.

Recipe for a Positive Learning Environment

Throughout this book I emphasize the importance of developing the social-emotional skills and 21st century life skills that learners will need to thrive in modern day society. "This includes

›tability, initiative, self-direction, communication, social and cross-cultural ɔductivity and accountability, and leadership and responsibility. The ability to .y and creatively with team members and classmates regardless of differences in yle is an essential 21st century life skill" (Trilling & Fadel, 2009, p. 80).

:ed the power of play and collaborative activities to teach these important life skills ee. This chapter offers a variety of practical strategies and collaborative group games that will engage learners of all ages in practicing and developing these essential skills WHILE building a positive, productive, and supportive learning environment. These methods are useful for teachers looking to build a collaborative learning community; advisory group leaders looking to enhance their curriculum and group work; college level student life staff looking to enhance first year and orientation programs; after school providers looking to add to their program repertoire; counselors leading family night events or "step up days;" principals leading staff building exercises; and trainers looking for engagement and group-building strategies.

In chapter four, I compared teaching and group facilitation to cooking. When you first start cooking you follow the tried-and-true recipes from family, a friend, or a cookbook. As your skills increase, you start to improvise and vary the ingredients. This chapter offers "recipes" for building executive function, practicing social-emotional learning, and cultivating 21st century skills. These recipes leave a great deal of room for adaptation, experimentation, and adjustment of ingredients to fit different groups and learning situations. Though there is room for a great deal of creativity and innovation, it is important to remember that there are key ingredients from the science of teaching, including educational neuroscience, educational psychology, and social-emotional development research that will help us maximize these learning outcomes with our groups.

Key Ingredients for Facilitating Social-Emotional Skills and Community-Building Activities

The Experience Starts the Moment Learners Enter the Room (or Even Before)

The first few minutes of a class or group session is your opportunity to draw learners into a positive learning experience and increase engagement. Welcome learners into the classroom/meeting space and focus their attention on the tasks at hand. Reflective practice to increase meaning, retention and transfer can begin in these very first moments. For ideas on creating a hook, see chapter five.

Take Time Up-Front to Build Your Group

Time spent at the beginning of your program or school year building relationships, ownership, and reflective skills pays off later in many ways. Use appropriate beginning activities that build trust in incremental ways. This is the foundation that will allow you to engage learners in more challenging activities, foster more responsibility and control over learning, and increase participant's ability to move learning forward.

Make Thoughtful Activity Choices

When choosing introductory activities, find those that build rapport, camaraderie, connections, shared understanding, commonalities and goals in an enjoyable and non-threatening way. Sequence and build healthy trust in your group.

Beware of the Icebreaker

An icebreaker is meant to "break the ice"—to build respect and a willingness to share that are essential to a productive and supportive learning community. When people are put "on the spot" too early in the group process, however, icebreakers can do the opposite of what was intended. Many people have negative connotations of team-building and ice-breaker activities because they have been put in situations where choice and control were taken away.

There is a fine line between a challenge that helps move learning forward and what the educational philosopher John Dewey (1938) would call a miseducative or potentially damaging experience. There are still educators who believe that stress can increase our ability to learn and be creative. There are icebreakers, games, and initiatives used in group work that were designed to quickly push people out of their comfort zones in order to initiate learning. Though we learn from novelty and from being challenged, it needs to be the right amount of challenge at the right time. Brain research validates Dewey's observation about miseducative experiences by showing that stress can inhibit learning.

Begin with partner sharing activities. This gives participants an opportunity to warm up by interacting with just one or two others at a time before sharing with the larger group. By building trust in this way, group members start to share and engage at their own pace and become more willing to push their comfort zones later on when it really matters.

Choice and Control are Essential

People learn best when they perceive a sense of control and have choice and ownership over their learning experiences. Think about creating opportunities that build this sense of choice and control from the very beginning of the program or school year. Empowering learners to set reasonable parameters around their participation creates an atmosphere of healthy trust and will increase involvement from reluctant participants.

Learn and Honor Names

Weave name activities and practice into introductory activities, starting with simple partner greetings and sharing before engaging participants in a whole group name activity. This chapter offers some effective methods for introducing and reinforcing names in a palatable way.

Establish and Reflect on Healthy Group Norms

Help group members create an environment where they feel responsible for themselves and each other. A supportive atmosphere will allow them to speak up when there is a breakdown in communication or an issue that affects the safety or potential experience of the group. It is ideal when the group takes that responsibility rather than the facilitator or teacher.

One benefit to using play and healthy competition in teaching and group facilitation is that the need to make decisions and work out conflict organically arises and groups have to find ways to figure out how to address these issues in order to move forward. When these natural conflicts or negative hurtful behaviors arise, they can become teachable moments if facilitated carefully. This is a perfect time to explore meaningful and relevant group norms. Educator-imposed ground rules or expectations should be put forward on day one, but participants should be increasingly involved in defining group norms as they move forward and encounter challenging situations together as a group (see page 50).

Consider the Physical Environment and Time Parameters

I prefer facilitating activities in the classroom or boardroom space rather than taking groups to the gym, cafeteria or outside. I find learners are better able to transfer the lesson to what goes on day-to-day in the classroom or boardroom when they experience the activity and reflection in that space. Many, if not most, of these activities require some movement of desks/furniture in the classroom. Reflective conversations are enhanced when the whole group can see each other in a circle. I find that students are motivated to jump in and help move furniture because they are excited about the novel activities coming up. Giving students responsibility for adjusting the room for interactive activities becomes a habit and does not take up much time (see chapter 4).

Of course it is effective to move the group to another space at times. I am fortunate that the schools and programs where I currently spend most of my teaching time have easy access to the outdoors, and we take advantage of it! Be aware of the varying physical needs of the participants in your group and adjust the space accordingly.

Reflect and Check In

Reflection brings learning to life. Regular check-ins create a forum or opportunity for participants to share what is working, what they need from each other, and for celebrating successes along the way. Reflective practice is best when it is a dynamic ongoing part of your lessons, not just something facilitated at the end of an experience or as a follow up. It plays an important part in the development and strengthening of neuronal pathways to enhance retention and the ability to recall and apply lessons (see page 14). Reflective activities are blended throughout this book (see chapter nine for more in-depth information on reflection tools and techniques).

Commonalities and Rapport-Building Activities

The first step in building a collaborative learning community is establishing rapport. An engaging opening activity designed to get the group interacting and sharing names, backgrounds and goals can maximize the outcomes of group process. Every facilitator has their ideal opening activities. Some of the best and most effective we learn from others or adapt from activities we already know. The activities in this section are those that have risen to the top of my list because they have choice and control built in and can incrementally build group rapport. Remember, it is not the activity itself, but how it is facilitated that makes the difference.

Don't be afraid to use an activity more than once for different purposes with your groups. The active review strategies presented in chapter six (i.e., Handshake Mingle, Commonalities Mingle, Concentric Circles) are great rapport builders because they start off sharing with partners and small groups before sharing with the large group. I recommend pulling from these when your purpose is community building and social-emotional skills development. Using these activities at the beginning of a program as rapport builders and then later in the program for reflection or academic content review increases the level of dialogue and participation. I find the group is familiar with the activity and therefore more willing to challenge themselves with the reflective or review questions.

Team Tally

Purpose/Focus: active engagement, transition, conversation starter, community building, fair play, communication, commonalities/different perspectives, reflection, context setting, academic review, formative assessment, differentiation, discussion technique, emotional connection to learning, empathy, introducing a topic, playful learning, peer interaction

Materials: team tally worksheet and a pen or pencil for each group

This group questionnaire serves as a "get to know you" or group celebration activity that involves friendly competition between teams. Group members talk through questions about their interests and experiences and come up with a cumulative score based on their group's commonalities. They learn about each other while engaging in healthy competition. Facilitators can involve their group in this self-directed activity during transition times when they need to attend to administrative tasks or setup.

I use variations of this activity with students, school faculty, adult team-building programs, family night programs, and after school programs. Some schools I work with use it in their advisory programs and then compete between advisories for the high score. This activity can keep after-school youth engaged during transitional activities such as gathering and snack time. It is also a great method for building rapport in lab groups or to kick off a group project. See the academic focus variation.

Facilitation Suggestions

- Divide participants into teams of 4 or 5, depending on the size of your group (try to make teams even).
- Have each group use a Team Tally sheet to come up with a cumulative score on each of the commonality areas.
- The team with the most points wins. If your groups are uneven, have them divide the total scores by number of team members.
- Have the group come to consensus about the parameters of some of the questions. Does a tank of fish only count as one pet? I have fun conversations here in rural Vermont about whether chickens or cows count as pets.
- Invite each group to share and celebrate their high scoring areas. The value in the activity is the sharing about commonalities.
- You can easily change the questions to fit the age group/population with which you are working or to focus on a specific theme or subject, such as the "back to school" or "science" team tallies shared here.

Outcomes/Reflections

Youth and adult groups enjoy this activity. The combination of sharing, competition, and collaboration draws in learners. I find that people make connections with peers outside their usual social circle. In a high school classroom, I had two students who usually did not get along well connect over a discussion about whether they should count the members of two households in the tally. Both students were struggling with recent divorces in their families and shuttling between two houses. This activity initiated a discussion and bond over similar circumstances that continued throughout the school year. In the rural area where I live, participants connect around outdoor activities. Adjust the questions to fit your group's personalities and goals.

Theme Based Variation

The advisory planning team that I work with at Mount Anthony Union High School decided to create a back-to-school variation of the team tally specifically focused on seniors. Molly Maish, Mike DiMaio, Amy Thivierge, and Mark Upright helped compile the Senior Advisory Team Tally to use during the first week of their advisory programs. The goal was to bring Senior Seminar (advisory) groups together, build camaraderie through shared experience, and initiate reflective conversations about career exploration and other senior specific planning.

Academic Focus Variation

Samantha Parker, a high school science teacher from Lincoln-Sudbury Regional High School, created a Science Team Tally to use in her physics class. Her goals were to build community among lab groups, introduce physics concepts, create context, and promote inquiry.

Resources/References: This activity was inspired by Rohnke (*Funn N' Games)* who put together a team-building worksheet focused on common teaching experiences for a program we facilitated together. He called it "Yeah Teachers." I have adapted the idea to groups of all ages.

Team Tally

Points / Questions

______ One point for each person living in your home.

______ One point for each pet in your household (fish only count as one pet).

______ One point for each person who has helped a family member with a chore or task in the last week.

______ One point for each team member who has ever fixed a car or bike.

______ One point for each team member who has ever helped paint a house.

______ One point for each team member who can name a new activity or sport they have tried this school year.

______ One point for each team member who has been on a pair of skis, snowshoes, or a snowboard or skateboard.

______ One point for each person who has volunteered to help others in their community in the last year.

______ One point for each team member who has made something themselves that they use day to day (i.e., clothing, furniture, journal)

______ One point for each team member who plays a team sport.

______ One point for each team member who has been to a concert this year.

______ One point for each musical instrument team members know how to play.

______ One point for each team member who can name a new activity, artistic endeavor, or sport they have tried this past year.

______ One point for each team member who has ever planted a garden.

______ One point for each team member who has been camping.

______ One point for each team member who has read a book for fun this month.

______ One point for each team member who has ever paddled a canoe.

______ One point for each button on all clothes worn by the entire team.

______ One point for each person who can name a time they have helped someone else during this school year.

______ One point for anyone who laughs at least once during an average school day.

TEAM TOTAL POINTS ________________

Senior Advisory Team Tally

Points / Questions

______ One point for each person who tried something new this past summer.

______ One point for each person who started a new job.

______ One point for each person who added to his or her savings account.

______ One point for each person who spent a night outside this summer.

______ One point for each person who slept past 9:00 a.m. this summer.

______ One point for each person who saw the sun rise this summer.

______ One point for each person who caught a fish this summer.

______ One point for each person who went to a concert this summer.

______ One point for each person who climbed a mountain.

______ One point for each person who is actually happy to be back in school.

______ One point for each person who enjoyed their summer reading.

______ One point for each person who is going to join a club or school sport this year.

______ One point for each television in your house.

______ One point for each four-legged pet that lives in your home.

______ One point for each person who went a day without watching TV this summer.

______ One point for each sibling of group members.

______ One point for each person who has welcomed someone new into our school.

______ One point for everyone who swam in a lake this summer.

______ One point for everyone who went to the ocean this summer.

______ One point for everyone who tried a new food this summer.

______ One point for everyone who went or will be going to a state or county fair.

______ One point for everyone who broke or lost his or her phone this summer.

TEAM TOTAL POINTS ________________

Ms. Parker's Science Team Tally

Points / Questions

______ One point for each person living in your home who wears eyeglasses/contacts.

______ One point for each magnet on your home refrigerator (estimate).

______ One point for each time this summer you hung a load of laundry outside to dry.

______ One point for each team member who has secretly been really interested in a science topic in class but didn't want the teacher to know.

______ One point for each team member who has secretly been really interested in a science topic in class but didn't want classmates to know.

______ One point for each science class you have taken.

______ One point for each team member who has used something learned in a math class in real life.

______ One point for each team member who has tutored a classmate.

______ One point for each calculator team members have with them now.

______ One point for each pen or pencil team members have with them now.

______ One point for each team member who has flipped a coin to make a decision.

______ One point for each team member who has rubbed a balloon on a piece of clothing then stuck it on a wall.

______ One point for each team member who has gotten a shock from a doorknob or car door.

______ One point for each team member who has tried to levitate something with his or her mind.

______ One point for each team member who has wondered why glass is see-through but walls are not.

______ One point for each team member who has wondered what it feels like to be weightless on a space station.

______ One point for each team member who has wondered what would happen if our Sun became a black hole.

TEAM TOTAL POINTS ________________

Have You Ever?

Purpose/Focus: playful learning, active engagement, social-emotional learning, communication, active listening, community building, commonalities/different perspectives, self-regulation, turn taking, reflection, academic review, differentiation, formative assessment, celebrations/appreciations of others

Materials: spot markers (yoga mats, carpet squares or bath mats can be cut for this purpose)

This activity is a well-known ice-breaker/get-to-know-you activity that is used in a variety of settings because it is fun, interactive, and easy to play. In chapter six, I shared Anyone Who (page 98), adapted from this activity as an active approach for reflection, academic review and formative assessment. Participants tend to buy into this adaptation of this activity because there is choice and control built into the game (i.e., participants can choose whether to move when a question is asked). It works best a little later in group process when the group has developed some comfort with each other and sharing.

Facilitation Suggestions

- Provide a spot marker for every person minus one in a circle on the floor.
- Then place a spot of a different color in the perimeter of the circle. This becomes the "hot spot." (Traditionally the hot spot was in the middle of the circle.)
- The person standing on the hot spot asks a get-to-know-you question such as "Have you ever climbed a mountain?"
- Then anyone who has climbed a mountain leaves their spot and tries to find a new one.
- Someone new ends up on the hot spot and asks another question of the group, sharing something about themselves and looking for commonalities with other group members.
- I add more opportunity for choice and control within the game by providing a buzzword such as "Bananas" for the person on the hot spot to use if they can't think of a question to ask. When this buzzword is used everyone has to move.

Facilitation Notes

Increase involvement in this activity by acknowledging that participants who are more introverted can sometimes find this game fear-inducing. As explained in chapter one, when some people are put on the spot to come up with a question they can experience enough of a stress overload to actually decrease cognitive engagement (Willis, 2010a, 2012).

Michelle Cummings of Training Wheels shared the idea of moving the hot spot out of the middle, and instead, providing an alternative color spot in the perimeter of the circle for the question-asking spot. Since then, I have facilitated the game without the middle spot and seen participation and buy-in increase. It also makes for much better communication as the person asking the question does not have his/her back to half the group.

When I first learned this game, facilitators often added a rule that you could not go to the spot next to you. After discussions with participants who shared their negative experiences with this game, I stopped adding this rule. I found this increases participation, as people were willing to try taking one step over. If that works okay, they might be willing to push themselves further next time, and so on.

High school, college, and adult educators have voiced concerns that this game can become inappropriate with group members asking questions that are offensive or inappropriate for the setting. When discussing if they should make a statement about "keeping it clean" at the beginning of the game, I explain that with appropriate sequencing (e.g., facilitating this after the group has established some comfort with the process), this is less likely to happen. If it does happen, that is when you stop the game and create a group norm about what is appropriate. When working with adolescents, I find that making a statement like that up front can communicate a lack of trust or plant an idea that wasn't even there. I find kids keep it appropriate and safe because **they want to keep playing**!

Variations

A camp director once commented to me: "I watch groups play this game and everyone seems to love it, but I wonder, are they really finding out about commonalities and others' experiences, or are they just focused on getting themselves to the right spot." So we improvised and added the option for the person who lands on the hot spot to share about their reason for moving. It worked well and added depth and meaning to the game.

My friend and colleague, Cynthia Paris, sometimes changes this activity to "Have You Ever Wanted To..." or "Do You Remember When...." These variations fit nicely with the reflective variation I share in chapter nine.

chool students came up with the fun variation of adding different movements to the en the participants on the hot spot ask their question, they can add a movement (i.e., ne foot," "heel toe only," "skip"). This adds a lot of laughter and slows the game down a up members can process the questions. I use this variation with very young children.

Facilitation With Very Young Children

Very young children are still learning the norms and rules of play. In the original version of the game, I found many would hover around the questioning spot instead of running from it. Naming it the "hot spot" helps negate this. If they end up on the spot more than once, I simply let them know they need to ask for a volunteer to ask the question. This works well and teaches them the need for taking turns to keep a game interesting.

Resources/References: This game has many names and variations. I first learned Have you Ever? from Rohnke (*Quicksilver*, 1995). Patrick Torrey shared the idea of using a variation of Have You Ever? for reflection (*A Teachable Moment* by Cain et al.). My clients and students over the years helped me adapt this version to the reflective and academic review variation, Anyone Who?

The Importance of Names

A key ingredient for creating a positive environment is to help participants know and use each other's names correctly from the very beginning of the school year or program. Even in small school districts where educators might assume students already know each others' names, many actually don't or are not always using and honoring each others' names in a respectful way. Given names represent choice, thoughts, and feeling on the part of the individual's family members. Nicknames or new or shortened names represent the choice of the individuals themselves and their identity. Name activities help groups of all ages practice names and explore the concepts of connecting with others, honoring individual strengths and personalities, and showing respect for individualism, identity, and choice. The following are my tried-and-true activities for helping groups learn and practice names in a playful, non-threatening, and palatable way.

> This bestowal of name and identity is a kind of symbolic contract between society and the individual. The name differentiates the child from others; thus, society will be able to treat and deal with the child as someone with needs and feelings different from those of other people. Through the name, the individual becomes part of the history of the society, and, because of the name, his or her deeds will exist separate from the deeds of others.
>
> **—H. Edward Deluzain, 1996**

Handshake Mingle (with Names Emphasis)

Purpose/Focus: active engagement, names, social-emotional learning, community building, movement, peer interaction, communication, commonalities/different perspectives, respect, appropriate touch, reflection, academic review, conversation starter

This effective, multi-purpose activity is presented again here because it is one of the most effective name and rapport-building activities I have ever used. Many adult participants have told me that, because of the Handshake Mingle game, they finally learned the name of that person who they had been passing in the hallway. And, even months after the in-service day, they often greet each other with the handshake when passing.

Facilitation Suggestions

- Handshake Mingle is described on page 89 in chapter six as a reflection and academic review strategy.
- As you guide participants through the different partner greetings simply say, "Make sure you know your partner's name."

One high school student stated in a journal entry, "I am getting to know people better, and I can actually say 'Hi' to them in the hallway when before I didn't have a clue who they were. And there is meaning behind the hello instead of a hello that means absolutely nothing."

Name Meanings Activity

Purpose/Focus: community building, names, social-emotional learning, 21st century skills, respectful behavior, honoring individuals, celebrating differences, peer interaction, trust, communication

We can better remember a person's name when we have meaningful information associated with it. When participants already know each other's names, this activity can help them learn more about each other, and it can lead to discussion around the importance of honoring a person's name (i.e., pronouncing it correctly, using it appropriately, showing respect for each other). This activity works best after a few warm-up activities like the Handshake Mingle, when the group is more comfortable and willing to share.

Facilitation Suggestions

- Ask group members to line up silently by the number of letters in their name or preferred nickname.

- Next, have them form a circle and take turns sharing their name. Ask them to clarify the correct pronunciation and share "the story of their name" (i.e., what they know about its origin or meaning, whether it is a family name).
- Emphasize that they don't have to share a story if they don't have one to share. Most importantly this is an opportunity for students to take ownership and share with the group the CORRECT pronunciation or nickname, something that often gets overlooked or mistaken in school settings.

Outcomes/Reflections

Learners of all ages have shared they prefer this activity to other name games because they get to share information that is factual and they don't have to think of contrived additional information (i.e., a favorite fruit that begins with the same letter as their name). It even works with large classroom groups. Students seem to want to listen and share this kind of personal information. Even very young children know about their name origin and/or feel strongly about what they want to be called.

I used the Name Meanings activity with middle school students partway through the first semester. In every session, there were students who made corrections to their name, including a mispronunciation that both teachers and students had perpetuated for 8 weeks into the school year. Some students were being called by a shortened version of their name that they didn't like (i.e., "Nate" for Nathan). Teachers were surprised because they always gave students a chance to correct them at the beginning of the year.

A teacher who had been at her school for 25 years told a group of colleagues during an advisory training that she really preferred "Patricia" over "Patty," but somewhere early on in her tenure at the school a secretary called her Patty and it stuck. After the group discovered this, they worked to rectify this with their colleagues at school, honoring Patricia's request.

I worked with a group that included Native American participants who shared that from the 1950s through the 1970s they had to drop their given traditional name at boarding school and replace it with a Catholic name. This activity can stimulate conversation with family members and be tied to a family tree activity. It can lead to meaningful reflection and discussion about names and identity, culture and history.

One teacher said, "When I have students with the same name in my class, I assign them nicknames like Michael, Mike, Mike A. without asking their preference. I will never do that again."

I also learned from one group that there are no specific name signs in ASL for common English names. Name signs are individualized for each person using descriptive name signs based on something about the individual's physical attributes, personality, hobbies, or past experiences.

Resources/References: I learned this activity from a group of teachers in Laconia, NH. It is also in my book, *Tips & Tools: The Art of Experiential Group Facilitation.*

Name Roulette

Purpose/Focus: playful learning, names, peer interaction, fair play

Material: a spot marker to designate the "turn around" point

This active, playful group memory game is a great way to learn and reinforce names. This activity will encourage participants to study up on each other's names. Give teams a few minutes at the start of the game to check in with each other and review names together—friendly competition can be a great motivator!

Facilitation Suggestions

- Divide participants into two groups and have them form side-by-side circles.
- Place an object between the two circles that acts as a marker.
- Have both circles of participants shuffle left or right while facing the center of their circles—no looking over shoulders.
- When you say "Stop," the two participants who are at the marker turn around and name the person they are facing.
- Whoever names the other person first captures that person onto their team, and the captive must join that circle.

Resources/References: I learned this kinesthetic activity from Karl Rohnke.

Peek-a-Who Names Review

Purpose/Focus: community building, name practice, empathy, playful learning, taking turns, communication, honoring individuals, commonalities/different perspectives, peer interaction, fair play

Material: a blanket, sheet or tarp

Originally, I used this activity with elementary students, but I'm finding that adults and adolescents also enjoy the playful, hiding aspect of the game. Brain research shows that people of all ages like to hide, play, guess, and predict. As with Name Roulette, when presented with the challenge, participants will study up on each other's names. Give team members a few minutes at the start to review each other's names together.

Facilitation Suggestions

- Divide the group into two teams using the Which One? activity on page 65.
- Have two facilitators or group members hold up a large blanket or tarp vertically.
- Ask teams to stand on opposite sides of the blanket, hidden from the view of the other team.
- Each team then chooses one person to move up close to the blanket (for young students, taking turns can be an important aspect of the game).
- On the count of three, the blanket drops and the two participants race to name each other.
- The winner captures the other person onto their team.

Variation

Later in this chapter, I share an appreciations version of this activity for building positive group norms, practicing positive descriptive language, reflection, and celebrating group members' strengths.

Resources/References: This game is a New Games Foundation favorite and was later coined Peek-a-Who by Rohnke and Butler in *Quicksilver*. The Complements Version is from *Tips & Tools* by Stanchfield.

Trade and Share Activity with Names Focus

Postcards, image cards, and objects are powerful ways to spark introductory conversations. In chapter six (page 91), the Trade and Share activity is used as a way to share reflective information in class. This activity is also a useful community-building strategy and an engaging method for learning and practicing group member's names. Many advisory group leaders I work with now use the activity for that purpose. See the Strengths Variation on page_.

Pass It To

Purpose/Focus: name reinforcer, playful learning, peer interaction, community building, self-regulation, taking turns

Materials: a rubber chicken, pool noodle, or other soft prop that can be used to gently tag participants on the feet

An active name review game for when the group has built some comfort with each other and the group process, developed some positive group norms, and had time to practice names.

Facilitation Suggestions

- Ask the group to form a circle. This can be sitting or standing, depending on the space.
- Place one person in the middle of the circle with a soft prop (something appropriate for tagging). I use a rubber chicken.
- The person in the middle is going to be the "tagger"—gently tagging other participants' knees or feet with the soft object.
- Someone in the circle starts off with, "I pass it to Johnny." Johnny responds, "I pass it to Sue." The person in the middle tries to tag Johnny or Sue before they can "pass it to" someone else.
- As with any "person in the middle" game, create an opportunity for the middle person to have an out (i.e., they can pass their position off to anyone at any time).

Resources/References: Rohnke, in *Bottomless Bag Again* (1997), calls his variation of this activity Whaumpum and attributes it to Lee Gillis.

Name Guess (My short and silly name trick activity.)

Purpose/Focus: names, playful learning, community building, active listening

In elementary and middle school classrooms, I sometimes play Name Guess—a strategy made up on the fly many years ago—to learn everyone's name.

Facilitation Suggestions

- One student is the guesser and turns his or her back to the group.
- The group comes up with a phrase that each group member repeats (i.e., "Keep Calm and Carry On," or "I am what I am.") Choose a pointer to indicate who speaks when.
- The guesser has to recognize other students by their voices–students love it! (Obviously this would not be appropriate for groups where there may be students who are deaf or hard of hearing; however, I bet your students could help you come up with a variation!).

Collaborative Problem-Solving and Communication Activities

One of the most important lifelong skills to develop is the ability to communicate clearly and effectively. This includes both expressing oneself and listening to others. Groups of all kinds can become more effective when they explore how they communicate with each other and the barriers to communication that exist in their groups. Collaborative problem-solving challenges help groups explore creativity, innovative thinking, experimentation and the value of having differing skills and perspectives in a group. Leadership skills can be practiced, including knowing when and who to follow and when to step in and when to step back. This section offers some of my favorite group challenges that can be adapted to a variety of age groups and educational settings.

Tin Can Pass

Purpose/Focus: communication, collaboration, community building, problem-solving, patience, creativity/imagination, innovation, active listening, goal setting, giving and receiving directions, trust, social-emotional learning, 21st century skills, peer interaction, playful learning, planning, leadership/followership, experimentation

Materials: a number 10 tin can, found in the recycle bin of most school kitchens

This problem-solving, communication, and group-building activity is elegantly simple as far as materials and setup. It requires encouragement and coordination between group members. It is a great introduction to group problem-solving that works well with all ages and can be adapted for small or large group sizes.

Facilitation Suggestions

- Have the group sit in a circle (floor, chairs, or both are fine).
- Level 1: Challenge group members to pass the can around the circle using only their feet (specify that this means below the ankle) without dropping it.
- Level 2: When the group is successful with level one, introduce level two by placing a tennis ball or other prop in the can, which often forces the group to change its strategy.

- Level 3: For advanced challenge, have group members close their eyes and pass the can. They can open their eyes when they have passed the can to help verbally guide others in the group. (As with any eyes-closed challenge, group members choose what's comfortable for them.)

Facilitation Notes

If the can is dropped, have participants decide what to do, or suggest that they send it back one person or restart right there and continue the activity. Sending the can all the way back to the beginning might become too frustrating for some groups. It is important to be aware of the necessary balance between challenge and achievable success experiences.

For larger groups, divide the group in half and have two circles take on the challenge. To keep everyone in one circle, use two cans and send one to the left and one to the right. This keeps more group members engaged and adds the challenge of crossing the cans.

If a group member doesn't want to take on a challenge, he or she can be the group reporter and write down what they hear and observe. This gives the group some interesting and usually amusing material for a reflection discussion.

Outcomes/Reflections

I often follow this activity with the Images/Postcard activity on page 71. I lay out the collection of postcards and ask the group to agree on one card that best represents what they achieved together in the Tin Can Pass activity. This activity is a useful introduction to problem-solving, and using the postcards in this fashion, introduces the group to metaphoric reflection. As mentioned in chapter nine, postcards and other object and image methods are effective reflection tools because participants can attach their thoughts to an object or image that can be touched and shown to a group.

Resources/References: I learned this activity from Karl Rohnke many years ago. My students at Steven's Point, Wisconsin added the level variations.

Rock-Paper-Scissors Tag

Purpose/Focus: collaboration, social-emotional learning, 21st century skills, communication, conflict resolution, consensus, compromise, decision-making, listening, leadership/followership, dealing with frustration, fair play, community building, playful learning, peer interactions

Materials: spot markers or cones to delineate the home base or "safe" area

This tag game that is so much more is perfect for team-building, advisory groups, after school, field days, and family night. It has become a favorite of mine because it reinforces a method of practicing consensus youth still use today, so it really seems to hit home when we play it with a purpose. It can be adapted for small classroom spaces by adding a rule that players move in a heel-to-toe walk, which slows everyone down to an appropriate speed for a small space.

Facilitation Suggestions

- Divide the group into two teams. See the Which One? activity on page 65.
- Explain Rock-Paper-Scissors to the group. Most participants are familiar with it as a way to make decisions in social and play situations without a mediator. Rock breaks scissors. Scissors cuts paper. Paper covers rock.
- Use poly spots or other markers to set up boundaries. Delineate a middle "face-off" line and two lines on either end of the playing area that represent each team's "safe" area.
- Instruct teams to huddle up in their safe area and decide as a team whether they are going to be rock, paper, or scissors. Emphasize that they can only be one of these things. This is where leadership skills, conflict-resolution, and compromise come in as some groups struggle with working together to agree on a sign. Sometimes I have teams choose a backup option in case of a tie.
- Remind players that this is Rock-Paper-Scissors Tag only, not cement, glue, dynamite, or other various alternatives (kids and young adults will know what you are talking about).
- After both teams have come to a decision, have them come to the middle face-off area.
- On the count of three, teams show their signs. The winning team chases the opposing team.
- If a participant is tagged before reaching the safe area, she is captured and joins the tagging team.

Facilitation Notes

People of all ages get caught up in the fun and excitement of this game and will usually eagerly participate in a number of rounds. Sometimes groups will use strategy, luck, or some kind of consensus to end the game. Each group's approach is different and can lead to interesting reflection discussions on decision-making, group communication, and consensus.

Outcomes/Reflections

In a high school challenge course class, we used Rock-Paper-Scissors Tag as a warm-up during our first week. When we read student's journal statements, we were surprised by their reactions to what we thought was purely a fun icebreaker. They journaled more about the dynamics of group decision-making, working with students from different grades and cliques, and the challenges of coming to group consensus on the rock-paper-scissors sign than they did about their climbing-wall experience and other activities we thought were more emotionally charged.

That same week, I used Rock-Paper-Scissors Tag with 4th and 5th graders as a communication-builder and a warm-up to another problem-solving activity I had planned. One of the teams repeatedly came to middle with all three signs. Clearly they had not agreed on a group strategy. I stopped the action and asked the group what was happening. This led to an in-depth discussion by the students around decision-making and leadership, and the difference between a democracy and a dictatorship (topics they were studying in social studies). They also tied this experience to a recent school election and how they wanted decisions made in their classroom. As evidenced by the discussion, this activity became much more than just an icebreaker!

Resources/References: I was first introduced to this activity through The New Games Foundation, who credits John O'Connell as the creator (Fleugelman, 1976). In their second book it evolved to Giants, Wizards, and Elves (New Games Foundation, 1981).

Community Circles

Purpose/Focus: social-emotional learning, 21st century skills, compromise, conflict resolution, community building, collaboration, creativity, active listening, critical thinking, following directions, innovation, problem-solving, communication, leadership, responsibility, boundaries, appropriate touch, big picture perspective, group norms, playful learning, peer interaction

Materials: rope circles of various sizes, one for each participant

This activity is a great metaphor that can be used in many ways. It requires rope circles and space in which to move around, so you will need some preparation. With school groups, I often facilitate this when it is easy to step outside the classroom.

Facilitation Suggestions

- Give everyone a rope circle in which to stand. Explain, "This game has just one important rule: everyone's feet need to end up inside the perimeter of a circle."

- Then tell everyone to switch circles.
- When participants begin moving to new circles, they immediately start asking questions like, "Can we walk between the circles?" Just remind them of the one important rule, "Everyone's feet need to end up in the perimeter of a circle."
- Go around the group, checking to see if feet are in. Then before you say "switch" again, take away a circle. Continue this process.
- The group will discover how to share circles—sitting down and putting their feet in the circle is a creative solution that stays within the parameters of the challenge.
- After the group has solved the challenge, and while they are already together in the circle, I ask them to get comfortable and share their reactions to the activity.
- I have also followed up in class with reflective writing prompts:
  - › What lessons did you learn from this activity?
  - › Reflect on our discussion. How can we take the lessons learned from this activity back to the classroom/school to improve our community?

Outcomes/Reflections

This activity usually generates reflective group discussion without much prompting. Participants share lessons learned about thinking outside the box, listening carefully to parameters, the importance of not making limiting assumptions, the value of group think, knowing when to follow and support a good idea, and the need to share and work with peers. Sometimes students will reflect on the limitations they impose on themselves at school such as who to sit with at lunch, what defines popularity, what to wear, etc.

Resources/References: Rohnke and Butler share this activity also known as Mergers or Star Wars in their book, *Quicksilver*. It was commonly played in the camping and adventure education world many years ago. Now I find it is new again!

Telegraph

Purpose/Focus: playful learning, social-emotional learning, 21st century skills, communication, focus, executive function, self-regulation, trust, gossip/rumors, appropriate touch, fair play, collaboration

Materials: a quarter, chairs in two rows (or space for a group to sit on the floor in two rows), a fun prop such as a rubber chicken or stuffed animal

Participants of all ages love the playful and competitive nature of this game. It is an effective way to engage groups in a discussion about communication and miscommunication, and one of my favorite ways to explore the concepts of thinking before acting, how one's actions affect others, gossip, and rumors. I share this activity in chapter six as a method for exploring probability in math. The game requires appropriate touch, which can become another social-skills lesson in itself for many groups.

Facilitation Suggestions

- Create two equal teams and have participants sit in chairs or on the floor in two rows facing each other.
- Place a novel object such as a rubber chicken precisely between the last people in each row.
- Ask team members to hold hands.
- Have everyone close his or her eyes except the first person in each row.
- The front persons are instructed to watch the facilitator flip a coin.
- If heads come up, the front persons send a squeeze hand-to-hand down the line. If tails, they don't! When group members feel a squeeze, they pass it down the line.
- When the squeeze reaches the last person in line, they grab the object. Whichever team grabs the object first wins that round. BUT if they squeeze on tails, they lose a point.
- Have students predict how many tosses will be heads and how many tails.
- Rotate front to back after each turn. Team switching can help eliminate negativity around competition. Ever so often have "anyone with a summer birthday" or "anyone wearing green" switch to the other team. This adds a little fun and controlled chaos, potentially evening out the teams, and helping the group let go of score keeping.

The **hand holding** aspect of this game can become an important reflection point about community building, trust and rapport in this activity. The desire to play the game usually outweighs any ambivalence about hand holding, but proper sequencing is key to the success of an activity like this.

- With uneven groups, the odd person can rotate in and out of the "judge" position, determining who grabbed the object first, or let them be the coin tosser.
- I often stop the game a few times to discuss the social-emotional skills lessons that inevitably arise with the "miscommunications" that occur. This game is a great metaphor for what happens in school regarding passing along incorrect information, acting before thinking, gossip, etc.

Outcomes/Reflections

This activity serves as a great metaphor for reflective discussion. Take advantage of the teachable moments by pausing the game regularly to talk about ways to improve communication and self-regulation in real life situations at school, in the workplace, and in other areas of life.

Resources/References: This is an adapted version of the game found in Jackson's *Activities That Teach* book, 1995.

"Follow Me"

Purpose/Focus: impact of our actions, leadership/followership, peer pressure, collaboration, group norms, discussion prompt, quick energizer, transition, 21st century skills, responsibility

This quick and simple activity initiates reflection on how our actions influence others and the impact of positive leadership. I use this when the focus of a lesson is on leadership or peer influence. It can also initiate discussion with a group of educators and/or counselors who are wondering whether their hard work is actually impacting others.

Facilitation Suggestions

- Ask the group to stand in a circle.
- Have everyone silently and secretly choose a person in the group whose actions they are going to follow.
- Ask everyone to close his or her eyes and strike a pose (simple poses such as hands on head, squatting, hands on hips, etc.).
- Ask everyone to focus on the person they agreed to follow and do exactly what they do.
- The group will go through a series of poses as the followers follow their leader. In most cases,

everyone eventually ends up in the same position, unless there is a "closed loop" of the same two people following each other.

Outcomes/Reflections

Participants are usually quite surprised to find everyone striking the same pose after a few moments. This demonstrates how people influence each other and how one person's actions can impact a larger group if people choose to follow. Possible reflection questions include: What did you notice during the activity? Were you surprised by the outcome? What real-life situation does this remind you of? Who do you choose to follow in your day-to-day life at school, work, and home? Do your actions make a difference? What "closed loops" in communication sometimes occur in our lives?

Resources/References: I learned this activity from Karl Rohnke.

Communication Breakdown

Purpose/Focus: social-emotional learning, 21st century skills, communication, frustration, giving/receiving directions, active listening, collaboration, community building, trust, leadership/followership

Materials: space to move around, various props such as stuffed animals or unbreakable items found in the classroom or board room that can be easily moved

This activity involves multiple "lines of communication" with participants who are verbally and visually limited. It is a great challenge for giving, receiving and interpreting directions.

Facilitation Suggestions

- Set up three lines of communication: line #1 is four participants at the front of the room (non-verbal directors), line #2 is a single person facing them (verbal director), and line #3 is a third person facing the back of the verbal director (the doer). The doer is challenged to complete a simple task with eyes closed, following the directions of the others.

- › Line #1/Non-Verbal Directors: The four participants at the front of the room can see the task at hand and all of the participants involved, but they cannot talk. They communicate non-verbally with the verbal director in line #2.
- › Line #2/Verbal Director: A single person faces the group of four non-verbal directors. This verbal director watches the first line of four participants with her back toward the doer, who is placed behind her. This verbal director can talk, but cannot see what the doer is doing.
- › Line #3/Doer: The doer has his eyes closed and stands a few feet behind the verbal director. This person will complete a simple task based on the directions of the first line of communication that will be interpreted by the verbal director in line #2.

- The rest of the group comes up with a simple task using basic props (e.g., moving a ball onto a chair or into a can, moving a book from a chair to inside a desk).
- The group not involved in giving directions is responsible for spotting the doer as he completes the task so he doesn't bump into anything.
- The task is shown to the non-verbal directors in line #1. They then try to non-verbally communicate the task to the person in line #2. The person in line #2 then explains the task verbally to the doer in line #3.
- The rest of the group spots person #3, but they do not otherwise interfere.

Outcomes/Reflections

This activity generates laughter and meaningful discussion around communication (and communication breakdowns) in the workplace or classroom.

Resources/References: This is a variation of a community-building challenge I learned from my colleague Johanna Liskowsky-Doak. There are many Lines of Communication activities used by various educators in the team-building field.

Zoom for Community Building, Communication and Problem-Solving

Purpose/Focus: social-emotional learning, 21st century skills, community building, problem-solving, communication, leadership/followership, managing frustration, patience, collaboration, different perspectives, higher order/critical thinking, active listening, big picture perspective, descriptive language, executive function, focus, sequencing, peer interactions, persistence

The *Zoom* activity that uses the picture book illustrated by artist Istvan Banyai (1995) is an excellent problem-solving and communication challenge. I list this activity here as well as in chapter six (see page 115) because of its strength both as a social-emotional skills-building tool and an excellent example of an active approach to teaching and reviewing academic or training content. Then again on page 47, I share an example of using this activity with school faculty as a way to initiate meaningful reflection on communication and collaboration between colleagues, interdependence, and understanding varying perspectives.

Back Art

Purpose/Focus: social-emotional learning, 21st century skills, communication, communicating salient points, focus, playful learning, fair play, appropriate touch, community building

Material: simple line drawings (i.e., star, house, cat, heart), blank paper, markers

This activity generates laughter and meaningful discussion around communication (and communication break downs) in the classroom or workplace.

Facilitation Suggestions

- Divide your group into two teams (large groups can be divided into three to four teams).
- Have the participants line up and face forward.
- Place blank pieces of paper and markers at the front of each line.
- Show the participants at the back of each line a drawing.
- Ask them to use their finger to draw what they see on the back of the person in front of them.
- That person than draws what they felt on the back of the person in front of them and so on.
- When the person at the front receives the "message" on their back, they draw what they felt on the piece of paper in front of them.
- Have the teams compare results. This will spark conversation on communication and miscommunication. The group can reflect on questions such as: Which team was fastest? Which was most accurate? What happened along the line? How is this like real life?

Resources/References: This well known activity is written up in activity books such as *Games for Teachers* (1999) by Cavert and Frank.

Back-to-Back Drawing

Purpose/Focus: social-emotional learning, 21st century skills, communication, patience, focus, playful learning, peer interaction, giving and receiving directions, leadership/followership

Material: two different drawings (I use connected geometric shapes and lines), a pen or pencil, two pieces of paper

Facilitation Suggestions:

- Divide the group into pairs.
- Have the partners sit back to back.
- One partner describes his/her drawing to the other partner so they can draw it without seeing it.
- When the partners finish, they compare the original with the partner drawing.
- Switch roles, using the second drawing.

Outcomes/Reflections

This is another way to initiate conversations about giving and receiving directions, the challenge of communication, describing salient points, listening and working together. I like to use it before beginning a group project.

Reference: This well known activity is also written up along with some great drawing templates in Jackson's *Activities That Teach* (1995).

Fill the Crate Challenge

Purpose/Focus: social-emotional learning, 21st century skills, communication, collaboration, problem-solving, creativity/imagination, innovation, using resources, planning, persistence, community building, critical/higher order thinking, leadership/followership, patience/managing frustration, peer interaction

Materials: milk crate full of soft objects of various shapes, sizes and weight; large perimeter rope; and 2nd perimeter rope for a planning circle

This favorite problem-solving challenge has many solutions, each unique to the group that takes it on. It can reveal a lot about a group's dynamics and working style, and it offers great food-for-thought to reflect on roles, communication, and collaboration methods.

Facilitation Suggestions

- Create a large perimeter circle on the ground.
- Empty the crate of throwable objects outside the perimeter.
- Place the empty crate in the middle of the circle.
- Challenge the group to get the balls/objects into the crate without stepping into the circle or moving the rope boundary.

Outcomes/Reflections

With some groups, I add a "no talking" rule. Younger groups will have a more difficult time with this, and you likely want them to practice dialogue and problem-solving. An option is to create another circle away from the problem where the group can talk and plan together.

This game can lead to great discussion. I often ask group members to each pick a metaphoric object that represents the role they took on during the activity (see chapter nine). This usually leads to a rich discussion about group dynamics, individuals' roles, and approaches to group decision-making and planning.

Resources/References: There is additional information on this activity and examples from group work in *Tips & Tools* by Stanchfield. This activity was adapted from a similar challenge presented by Jim Schoel at a conference in 1997 and has evolved through experimentation over the years.

Turnstile

Purpose/Focus: social-emotional learning, 21st century skills, communication, problem-solving, collaboration, taking risks, planning, using resources, community building, followership/leadership, playful learning, peer interaction, trust, persistence, gymnasium/field game

Materials: large jump rope (25-40 feet)

The whole group is challenged to cross to the other side of a turning rope without touching it. I sometimes create a fun story line to go with this activity. For example, they are a group of scientists trying to leave a space station, and a dangling pipe from the cooling system is blocking their way out. They want to get everyone through and back to earth without being hit.

Facilitation Suggestions

- Facilitators or participants turn the rope at a slow rate.
- Group members are challenged to see if they can all get through the spinning rope from one side to the other without being touched by the rope.

- There are a number of creative solutions, including carefully observing the upswing of the rope and yelling "go," with the more confident runners helping others. Sometimes the group finds an arc created very close to the rope turners that they can step or crawl under.
- Step up the challenge by having the group complete the task without an empty rotation—which means they either have to keep someone jumping the whole time till everyone passes through or run through consecutively without an empty rope.

Outcomes/Reflections

This is a great large group, problem-solving challenge that can lead to meaningful reflection around creative thinking, helping and encouraging others, taking risks with support from the group, and the value of trying more than one way to solve a problem.

Resources/References: This game was popular when I first started working on challenge courses years ago. The earliest reference I found for The Turnstile is in Rohnke's *Silver Bullets*.

Turning Over a New Leaf aka Magic Carpet

Purpose/Focus: social-emotional learning, 21st century skills, communication, collaboration, creative thinking, problem-solving, trust, spatial relations, innovation, experimentation, appropriate touch, planning, persistence, patience, leadership/followership

Materials: either a large tarp/blanket (approximately 6 x 10 feet for up to 15 participants) or large pieces of butcher paper, depending on the variation you are using

This problem-solving task involves close physical proximity, so it is best sequenced later in a group's experience together. It requires clear communication and innovative thinking. There are a few variations listed below.

Facilitation Suggestions

- Spread out the tarp and invite group members to stand on it.
- Challenge the group to flip the tarp/blanket over and stand on the other side, without stepping off of it in the process. Every participant must be in contact with the tarp at all times.

Variations

This activity can be especially challenging using large pieces of butcher paper instead of the tarp.

A "Magic Carpet" variation has the group move the tarp across the room without stepping off of it.

Outcomes/Reflections

Both methods require the group to think differently and inspire many creative solutions. I use this to facilitate conversations around collaboration when participants are starting or are in the midst of a group project.

Resources/References: Variations of this activity can be found in *Activities That Teach* by Jackson, 1995 and *Teamwork and Teamplay* by Cain and Joliff.

Newspaper Bridges

Purpose/Focus: social-emotional learning, 21st century skills, collaboration, problem-solving, communication, using resources, innovation/experimentation, planning, peer interaction

Materials: newspapers and masking tape for each group and one large soup can or full water bottle

This challenge is easily facilitated and does not take much time. It is a useful way to initiate discussions on personality styles and the roles learners take in a group project.

Facilitation Suggestions

- Divide learners into groups of 4 to 5.
- Distribute a roll of masking tape and equal stacks of newspapers to each group.
- Challenge them to build a bridge out of newspapers that is strong enough to hold the soup can or water bottle and high enough that it can fit underneath it.

- Give them 15 to 20 minutes to complete the task.

Outcomes/Reflections

I use this activity for discussions around how group members can approach group projects. It can be a team-building check-in for groups who are working on an academic project together.

Resources/References: This was adapted from Jackson's *More Activities That Teach.*

Helium Hoop

Purpose/Focus: social-emotional learning, 21st century skills, communication, collaboration, problem-solving, focus, patience/managing frustration, synergy, persistence, leadership/followership

Material: hula hoops, one for each group of 6 to 8 participants

This is an interesting and surprisingly challenging activity. It is perfect for a group that has been together for a while and needs to reflect on how individuals affect the group without even being conscious of it. Many facilitators use a tent pole for this activity—I like using a hula hoop so that all the participants can face each other.

Facilitation Suggestions

- Ask participants to form circles of 6 to 8, facing each other. Be aware that the more people involved, the more difficult the challenge.
- Have them lift their arms from the elbow out and extend their index fingers.
- Rest a hula-hoop on their index fingers, and challenge them to try to lower the hoop to the ground without anyone losing contact with it or curling their fingers around the hoop.
- As groups finish, they can coach other groups and/or talk together about their experiences.

Outcomes/Reflections

Groups initially think this is an easy task; they are amazed at how the hoop moves when they don't think they are affecting it. Participants may get frustrated and blame others, which becomes a valuable teachable moment. It illustrates how we unconsciously affect others and that when we focus together and let go of blaming, we are more likely to succeed. It's an excellent activity to use with groups who are collaborating on a project.

Resources/References: This is a variation of Helium-Stick.

Saving Springfield

Purpose/Focus: social-emotional learning, 21st century skills, problem-solving, communication, managing frustration, leadership/followership, decision-making, collaboration, creativity/imagination, peer interaction, using resources, critical thinking, community building, giving/receiving directions

Material: large rope to form the perimeter; milk crate; small bucket of ping pong balls; and a toolkit containing four long pieces of rope, surgical tubing or bungee cord; blindfolds (see Facilitation Notes for other ideas)

This advanced problem-solving initiative is commonly known as Toxic Waste or Nuclear Reactor. I decided to blend in a little popular culture and call the activity Saving Springfield (in reference to the popular TV show, *The Simpsons,* [1989, Matt Groening, Fox Broadcasting Company]). Students immediately engage in the fun theme. The group is given the hypothetical task of shutting down a nuclear reactor core (a bucket of ping pong balls in the center of a circle).

Facilitation Suggestions

- Create a large perimeter with a rope. Set a milk crate upside down in the center with a small bucket full of ping pong balls on top. Place the toolkit outside the perimeter.
- State the group's challenge:
  You are a group of technicians who have been called in to save Springfield from imminent disaster. You must shut down the core of the nuclear reactor (represented by the can of ping pong balls). It must be moved out of the center of the circle. No one can step inside the circle, and the can needs to be gently placed outside the circle without spilling any of the contents.
- Point out the bag of equipment that is available to them (see materials). Tell them that anyone who comes in contact with the core of the reactor while using the equipment must wear protective eyewear (blindfolds or closed eyes)—creating another layer of communication challenge.
- Always give participants a choice to close their eyes instead of wearing a blindfold. If I use blindfolds, I put out clean, silly, colorful scraps of cloth or have them make their own.

Facilitation Notes

The way this is set up makes it an advanced problem-solving activity. I use it with middle school students later in the school year as a "culminating" group challenge to test and practice the skills we have been developing over the year. It requires a great deal of communication; group decision-making; leadership, compromise, collaboration and creative thinking.

Variation

There are many variations for this activity—my favorite involves red herrings (e.g., play dough, toilet paper, plastic horseshoe, rubber chicken) and equipment that allows at least a few different choices for how a group might approach the task.

Outcomes/Reflections

Students often struggle with this activity over a few different class periods. At one point, we were concerned that it was too frustrating and becoming a "miseducative" experience. Then we read student's language arts journal statements about the activity:

We worked well as a team, but some felt left out .... We learned that we should really encourage people to get involved. Once we all knew what worked and what didn't, we really started to come together.

I think we are learning that to get somewhere we need to be open to new ideas and work together and include others or you won't get where you want to be.

... sometimes it doesn't work the first time. You have to brainstorm ideas then pick one and try it. If that doesn't work then pick another and try it. If that doesn't work then just go back to the drawing board. I also think we should listen to each other more so people could understand the concept or idea more.

... when things that are hard—even if they seem impossible at one point, if you back away and prepare for it, it won't be a problem. The more you prepare for something hard, the easier it is.

... since so many different people are speaking at once and saying their ideas, we can't hear them unless we make a plan and LISTEN. So I think we are learning that to get somewhere we need to be open to new ideas and work together and include others or you won't get where you want to be.

When things are hard we tend to give up and get frustrated but learning how to deal with them will help us more in the future. I learned that if something is hard, I shouldn't give up. I should just keep trying till I succeed.

... you just need to keep pushing through things. You also might not get it the first few times around or you might not get it at all—and that is okay, if we really tried and learned something from each other.

From things that are hard we learn two things. The first is how to problem solve and the second is how to listen to others' ideas—even if they aren't our friends.

We learned that experience is more important than intelligence, that teamwork is more useful than popularity, and actions do more than words ever can.

You should use everything that is available to you to solve the problem. Try different strategies and not just the same one over and over.

The 8th graders also mentioned the activity in their graduation speeches.

Resources/References: I originally learned a variation of this activity as Toxic Waste from my professor Mike Gass at the University of New Hampshire more than 25 years ago. Over the years it has been written about with many variations and a variety of names. It is commonly used in the adventure education and corporate team-building settings. See *Tips & Tools* by Stanchfield for more information on blindfolds.

Appreciating Others and Positive Group Norms Activities

Activities in this section promote empathy, understanding differences, speaking up as a responsible citizen, celebrating strengths and positive contributions, and taking ownership for creating a productive and supportive learning community. The aim of these activities is to build a safe and supportive environment where learners can practice the social-emotional skills they will need to thrive in our ever-changing, modern-day society.

Ongoing reflection upon and practice of positive group norms is an integral part of making this happen. As we discussed in chapter four, page 50, group norms are the values and characteristics that exist in every group. Activities designed to reinforce positive group norms appear throughout this book to emphasize the need to look at developing these norms as an ongoing process not an event. Ideally it is part of the formative stages of a program and then, through ongoing practice and reflection, positive group norms come to fruition. The following activities use strengths-based approaches for reflecting on group norms throughout a program so that a group can continue to adjust and improve them.

Peek-A-Who Appreciations and Celebrations

Purpose/Focus: celebrations/appreciations of others, mindset, strengths, social-emotional learning, descriptive language, 21st century skills, listening, trust, reflection, group norms

Materials: blanket, sheet, or tarp

Earlier in this chapter, we explored the name variation of this game. This appreciation/celebration version of the game gives learners an opportunity to give and receive positive feedback from their peers, learn the difference between physical characteristics and personal strengths or qualities, and focus on the unique qualities and contributions of the individuals in their group.

Facilitation Suggestions

- In this variation, one participant from each team is chosen to sit with his/her back to the tarp. (Two facilitators or group members hold the tarp vertically between them so that they do not know who is on the other side of the tarp.)
- When the blanket is dropped on the count of three, the back-to-back pair remain with their backs to each other while their teammates describe the person from the other team, using only positive, non-physical descriptors (i.e., "She is very creative," "He is good in math," "She climbed Mt. Washington last week," "She was the leader of the last group challenge").
- The back-to-back pair tries to guess who is behind them based on this information.
- Sometimes in their excitement, the group gets too loud for the guessers to hear the descriptions. During one program a 6th grader suggested that each guesser pick three people to hear descriptors from, and it worked really well. Before each turn the describers were chosen by the guesser (making sure everyone had a chance to be a describer).

Facilitation Notes

Most of the time there are lots of memorable positive comments; though it takes practice for many groups to stop using physical descriptors. This compliments version is a great opportunity for students to celebrate the strengths and positive qualities of their peers and hear some nice things about themselves.

Outcomes/Reflections

It can also be a vehicle for practicing and discussing positive group norms. When something negative is said about a student during this activity (with proper sequencing it should be a rare occurrence), you know it is possibly happening in other situations, so this is a great opportunity to address it and initiate a conversation on group norms. This has happened to me twice over the years. Both times it led to powerful teachable moments:

> The first was a group of sophomores in high school that I had been working with since the 8th grade. It was an economically depressed area, and the difficult circumstances and negative attitudes got us off to a rough start. Over time and with a lot of patience, we made amazing progress in building a positive community and developing positive social-emotional skills and norms. When I used the Peek-a-Who Appreciations game, one of the boys described a girl on the other team as "not the smartest girl in our class." The group immediately called him on the inappropriate meanness of his comment (something they probably would not have been able to do when I first met them). He realized the implications of his comment and apologized publicly and again later on. We had a meaningful discussion about group norms and how our comments affect others. However, there was more to the story. The girl in question was very smart, but at that time she was acting very silly with the boys in her class. It seemed to me and the other teacher that the boy was commenting on her recent behavior. The rest of the day, the girl stepped up and showed everyone just how smart she was.
>
> The other teachable moment arose when a 7th grader described a fellow student as "He likes SpongeBob [cartoon character]." The boy got upset and said, "I don't like SpongeBob anymore, that was back in 5th grade." He was very serious about this, and the other students gave him their full attention. I said, "What are you asking of your classmates?" He said, "I want everyone to stop teasing me about SpongeBob!" I asked the students if this was something to which they could agree. They responded positively, and we continued the game. When I returned a week later and asked the students what they remembered about our last session, someone stated, "Johnny asked us to stop teasing him about SpongeBob, and we have." Their teacher nodded.

Variations

Use this as a reflection activity, asking group members to describe the person by his or her positive contributions to the class, group project or challenge.

Resources/References: This is a New Games Foundation favorite and was coined Peek-a-Who by Karl Rohnke and Butler (Quicksilver). The compliments version is from *Tips & Tools* by Stanchfield.

Appreciation Notes

Purpose/Focus: celebrating/appreciating others, descriptive language, reflection, strengths and positive qualities of others, trust, mindset, group norms, citizenship, social-emotional learning, 21st century skills

Materials: tape, washable thin point colored marker, and a blank paper for each participant

This is fun, non-threatening way to engage students in giving and receiving positive feedback and appreciating each other's qualities and contributions to the group. It works best with groups who know each other well and need some self-esteem boosters and opportunities to celebrate the strengths of group members.

Facilitation Suggestions

- Tape a blank piece of paper on the back of each participant (with permission, of course), and give everyone a thin washable marker.
- Invite group members to visit everyone in the group and write a thoughtful compliment or some positive feedback about that person on his or her paper. Explain that they are to focus on personality strengths or positive actions rather than physical characteristics. Give clear examples of what kind of compliments are appropriate (e.g., "I appreciate your willingness to speak out when you have concern," "You are very observant," "You make a great leader.").
- After the students have visited everyone and written their compliments, give them the opportunity to sit down and read their papers.
- Invite participants to form a circle and share at least one compliment from their list.

Outcomes/Reflections

With clear directions and parameters set at the start, most groups will readily engage and participate in a positive way. This has been especially effective with adolescents, who may lack self-esteem and don't give and receive positives very often.

Resources/References: This well-known group-building activity is also described in *Tips & Tools* by Stanchfield.

Postcard Strength Activity

Purpose/Focus: celebrating/appreciating others, strengths and positive qualities of others, mindset, descriptive language, metaphor/imagery, reflection, community building, communication, citizenship, self-expression, peer interactions, group norms, writing prompt

Materials: postcards, images or objects

In chapters five and nine, we explore how using postcards and metaphoric objects and images can spark introductory conversations and initiate meaningful reflection on learning experiences. This Postcard Strength activity uses this technique as a morale and rapport-building experience that strengthens the whole team. Use this activity with smaller groups that know each other and with participants who are comfortable with group process and sharing with the whole group.

Facilitation Suggestions

- Ask group members to pick a card that represents a personal strength, a positive quality they bring to the team, or a strength or unique perspective they use in their professional practice (teacher, engineer, nurse, daycare provider, etc.).
- Then have participants take turns holding up their postcards, and ask their colleagues or fellow students to guess why they chose it.
- After receiving feedback, ask participants to share how close the group was to guessing why they chose the card. This gives individuals an opportunity to express thoughts about their personal strengths and contributions—another important skill to practice.
- I often invite participants to write a note about these strengths on the postcard as a reminder and take it with them as a memento. When appropriate, I have also asked them to self-address the postcard, and I send to them at a future date. See chapter nine for more ideas around using postcard reminders.

Outcomes/Reflections

Celebration the personal strengths and contributions of colleagues and/or fellow students is something that doesn't happen enough in the work-place or at school. In my experience, people have regularly share thoughtful compliments and insights that go beyond the reasons the individual chose the card.

Resources/References: This idea was first suggested to me by a participant in my workshop at the Chewonki foundation. You can find postcards at yard sales and flea markets. Experiential Tools offers "Pick-A-Postcard" postcard sets for educators at www.experientialtools.com.

Trade and Share Strengths

Purpose/Focus: reflection, mindset, celebrating/appreciating others, community building, group norms, active listening, communication, descriptive language, metaphor/imagery, peer interactions

Materials: postcards, images or objects

The Trade and Share activity in chapter five works well with larger groups where participants can practice sharing with partners before having a large group discussion. Objects or image cards can be used for this strength's celebration version. I prefer the postcards, so participants can write a note to themselves on the back.

Facilitation Suggestions

- With this variation, ask participants to choose a card that represents a personal strength and then follow the Trade and Share sequence described on page 91 (2 or 3 trades).
- During the "report out," ask group members to share what they heard from others, as well as their own observations, about the person's strengths and contributions.
- I usually follow this with an invitation for them to write a note on the postcard to their future self as a reminder of their strengths and then send it to them at a future date.

Outcomes/Reflections

My colleague Scott Salway and I facilitated this activity with 8th graders and it worked beautifully. After sharing what they heard about the person's card, we asked the group for any additional thoughts about their fellow student's strengths. Four or five meaningful, insightful comments were added about every one of the students. We then asked the student whose card was talked about to add or clarify his or her reasons for choosing the card. This was a powerful experience for our group. The students did an amazing job of giving and receiving positive feedback about their peers. It was great to hear them practice descriptive language, share positive stories, and to see the faces of the students as they heard from their peers. This was a perfect lead-in to having them write a postcard to their 9th grade self in Language Arts class a few weeks later (see page 217).

Resources/References: You can find postcards at yard sales and flea markets. Experiential Tools offers "Pick-A-Postcard" postcard sets for educators at www.experientialtools.com.

Vintage Keys / "Key to Our Group's Success"

Purpose/Focus: group norms, reflection, discussion/writing prompt, community building, citizenship, social-emotional learning, 21st century skills, responsibility, strengths, mindset

Materials: keys, worksheet with prompts, writing utensils

For years I have been collecting vintage keys, and I often include them in my toolbox as one of many choices. Recently I've been using the keys for reflective activities. A group of social studies students had just finished brainstorming a student constitution with their teacher, identifying what behaviors were going to be part of their group during the upcoming school year. We followed up with some group-building activities, and right away students started breaking the norms they had just established. We sat down to reflect. I asked them if they just wrote down words they thought their teacher wanted to hear or if they really believed in them. They admitted to writing what they thought "should" be on the list rather than what they felt would really work. They blamed others for not being able to follow the norms but took little responsibility themselves. We spent some time talking about individual as well as group responsibility.

I wanted to follow up the next day with an activity to reinforce the meaningful conversation we had about individual responsibility to group norms. On my way home, I went to the hardware store and bought a bunch of key blanks—two each of assorted designs. I thought the keys would make a useful metaphor for "the key" to building a positive classroom community.

Facilitation Suggestions

- As students entered the classroom, we gave them a key and asked them to respond on paper to the following prompts:
  - › In your opinion what is the key to creating a caring and supportive community?
  - › What are you going to do to create a positive and supportive environment in our school?
- We then had them find their key partner and discuss their answers to these questions. This led to a whole class discussion on making the group norms a reality. I invited students to take their keys with them as a reminder. Most of the students tied the keys onto their backpacks or shoes.

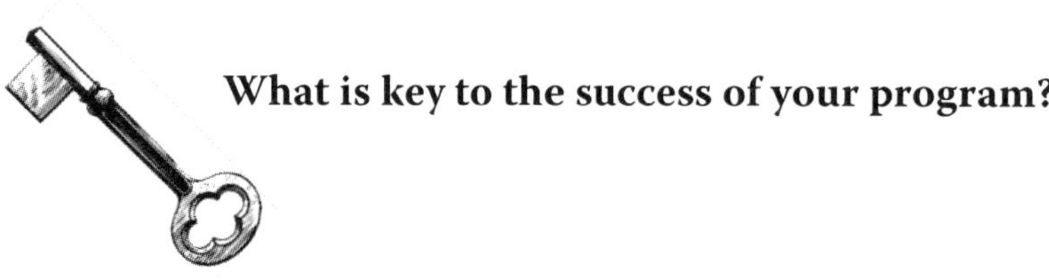

What is key to the success of your program?

What are you personally going to commit to doing to move our work forward together as a group?

In your opinion what is the key to creating a caring and supportive community?

What are you going to do to create a positive and supportive environment in our school?

Vintage Key Variation

This activity was also helpful in an adult, team-building session. After conflict arose during a problem-solving activity, I brought out the US activity (see page 50) to facilitate a discussion on group norms. It was an intense and difficult discussion in which group members called each other on behaviors in the workplace. The next morning as an entry task I brought my collection of vintage keys and asked everyone to choose a key they liked and respond on paper to the following questions:

› What is key to the success of your program?
› What are you personally going to commit to doing to move our work forward together? As a group?

Over the years, I have used this effectively with a number of teacher and student groups as a reflective prompt on their personal commitment to group development.

Resources/References: This naturally emerged from my toolbox activity. Matching key blanks can be purchased at any hardware store.

Community Puzzle Pieces "Strengths Mural"

Purpose/Focus: community building, strengths, mindset, reflection, celebrations/appreciation of others, group norms, creativity/imagination, social-emotional learning, 21st century skills, multiple pathways to learning

Materials: make your own giant puzzle or purchase blank puzzle pieces

At Twin Valley Middle School, I collaborated with the art teacher to create a giant mural of community puzzle pieces with the incoming 6th graders to represent their strengths and interests. When I shared this idea with a local elementary school guidance counselor, she began having all of the students along with the entire faculty and staff contribute a piece for a mural that would be placed in the entry way of their school as part of a bullying and harassment prevention program.

Facilitation Suggestions

- Puzzle pieces can be decorated using a variety of mediums to represent the individual student's roles, strengths, or

commitments to the group and how they relate to the whole.
- I use puzzle pieces with teachers to explore what they are willing to bring to our learning team, asking them to decorate their piece accordingly.
- Introduce this activity in one class session and give students time over a few sessions to complete their pieces. Some people would rather take time throughout the course or take the piece home.

When asked to work with a school administrative team on a new initiative for their district, I decided a **puzzle metaphor** would be helpful. Group members were asked to take their puzzle pieces home and decorate them in a way that defined their role and commitment to the group. One person cut her puzzle piece into five pieces that represented the many different aspects of her role in the group. Another participant just glued a giant pair of sunglasses to her piece to represent the "big picture" role she needed to take in the group. The puzzle led to some powerful discussions about each group member's strengths, fears, hopes, and level of commitment to the group. They kept the puzzle for a reference as they worked together that year.

Resources/References: A description of puzzle pieces used as a group norms activity appears in *Tips & Tools* by Stanchfield. Make your own puzzle pieces or purchase blank puzzle pieces from The Community Puzzle. They offer interchangeable pieces in bulk that can be combined to create very large puzzles for many group members (also available at training-wheels.com and experientialtools.com).

Draw Your School/Workplace

Purpose/Focus: social-emotional learning, 21st century skills, group norms, goal setting, community building, reflection, metaphor/imagery, multiple pathways, citizenship, collaboration, vision setting, responsibilities, creativity/imagination, communication, peer interactions

Materials: flip chart or butcher paper, markers/crayons for every person in the group

This activity focused on group norms initiates discussion and reflection on what ideal group behaviors look like, sound like, and feel like in order to create a shared understanding and vision around improving the school or workplace climate. It is a great follow up or lead in to the US list (page 50). This is especially effective when starting a new project or initiative or during times of transition such as grade level changes.

Facilitation Suggestions

- Divide the group into teams of 4 to 6 people.
- Ask each person to choose a marker or crayon (this increases involvement, by making sure everyone has a tool for drawing/writing if they are inspired).
- Ask group members to think about their school or workplace (any changes taking place).

Then have them work together to draw their ideal school or workplace.

- They can include some aspects of the physical plant, but the focus should be on how people interact (i.e., treat each other, talk to each other) within the space. What would be ideal? What does it sound like? What does it look like?
- Invite them to use words, symbols and drawings for their depiction. Ask that each group member be involved in some way.
- Allow 10 to 20 minutes, depending on the group.
- Then have each group stand or sit together and present their work to the rest of the group.

Outcomes/Reflections

This activity leads to reflection and discussion in both small and whole group settings about healthy group norms and acceptable behaviors. Time and time again, I find this conversation brings in the voices of people who don't always contribute to verbal only or list making oriented discussions. This is an option for what ideally should be a series of ongoing activities and reflections on group norms during a group's time together.

Resources/References: This group drawing variation was inspired by my work with middle schoolers at Twin Valley Middle School in Whitingham, VT.

Group Norms Sculpture

Purpose/Focus: group norms, reflection, social-emotional learning, 21st century skills, collaboration, citizenship, responsibilities, metaphor/imagery, creativity/imagination, multiple pathways to learning, communication, peer interactions, goal setting

Materials: play dough, clay, recycled objects, or other sculpting material

Previously we explored the power of metaphor and creative activities for reflecting upon and expressing important ideas. One of my favorite ways to use play dough is in group norms discussions.

Facilitation Suggestions

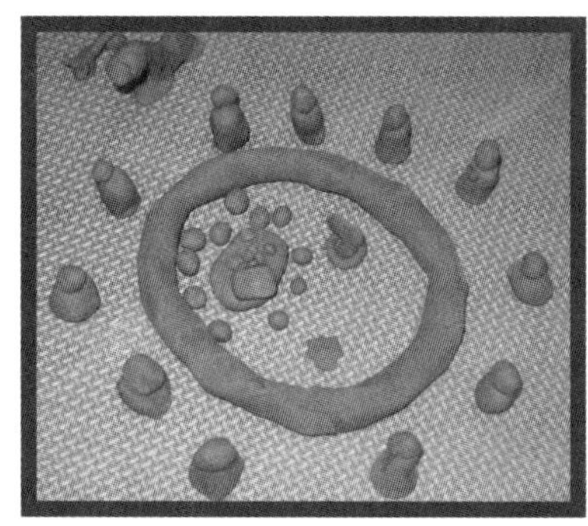

- Divide the group into small teams.
- Ask each group to work together to create a sculpture representing the most important behaviors, attitudes, and/or actions that a group needs to develop positive group norms.
- After 10 to 15 minutes ask each group to present and share about its sculpture.
- Small group sculptures can be linked to create one large group symbol of the group's goals.

Resources/References: This is a variation of Playdough Pictionary and the Draw Your School/ Workplace activities.

Group Symbols

Purpose/Focus: group norms, reflection, social-emotional learning, 21st century skills, collaboration, citizenship, responsibilities, metaphor/imagery, creativity/imagination, multiple pathways to learning, communication, peer interactions, goal setting

Materials: paper markers, sculpting materials, or found objects

It can be effective to revisit group norms throughout a group's development. Giving the group an opportunity to create or find a symbol for its strengths and goals is one way to facilitate this process. You can present reflective tools—such as postcards, objects, puzzle pieces, Chiji Cards (see resources)—that involve symbols of their experiences or successes as they progress through activities. Often groups will latch onto a symbol that can later become part of its group norms symbol.

Facilitation Suggestions

- Ask the group members to choose a symbol/ picture/card/object that represents the group's strengths.
- Invite the group members to create a drawing of this symbol

with words describing the strengths they each bring to the group, their commitment to the group, and the goals the group has defined for themselves.

Resources/References: *Tips & Tools* by Stanchfield explores group symbols on page 63.

Time Management, Planning, and Goal-Setting Activities

This series of activities initiates discussion and reflection on identifying priorities, responsibilities, and interests. It initiates exploration of topics such as the detriments of multi-tasking and prioritization of time.

Group Juggle Metaphor

Purpose/Focus: social-emotional learning, 21st century skills, collaboration, planning, communication, problem-solving, time management, coping, stress management, identifying resources, reflection, empathy, group norms, discussion prompt

Materials: group juggle worksheet, balls and other soft throwable objects, a large sheet of paper or white board, marker, paper and writing utensils for each participant

I often use this with school faculty, advisory groups, or health education classes to focus on coping with stress, planning, managing priorities, and identifying resources. I start by engaging the group in a variation of the well-known Group Juggle activity, in which the group works together to keep an increasing number of balls up in the air.

Facilitation Suggestions

- Have participants form a circle.
- A leader starts by sending one ball criss-crossing the circle to establish a tossing order that includes everyone in the group. Ask those who have not yet received the ball to put a hand up.
- When everyone has received the ball once, it comes back to the leader

Juggling Worksheet

who says something like, "Let's try that again using the same order."

- Gradually, the leader starts adding balls or other objects into the mix until it reaches a point of playful chaos.
- This is the perfect time to take a reflective break and ask participants to establish some goals and strategies: How many balls can they keep in the air at a time? How many times can they send the ball around the circle without dropping it?
- I often have the group practice setting realistic goals and making choices about what objects to keep in the game (which adds to the discussion later).
- After the participants meet their goal, have them brainstorm a list of the skills they used. Write these down, and keep this list in a visible place to refer back to later.
- I then use a simple Group Juggle Worksheet like the one on page 178 or have them draw a picture of a person (or team if you want to focus on a group task) juggling a bunch of objects with a few on the ground. The participants then label the objects with the responsibilities, anxieties, and activities they are juggling in real life.
- Group members then share and discuss what they identified as the "balls in the air" and the "dropped balls."
- Finally, I have group members look back to the list of strengths and skills they brainstormed during the group juggle activity and apply these skills to their real life juggling. I ask participants to identify the skills/resources/strategies they need to focus on together to better manage all they have to do.

Example Brainstorming List:

- Focus
- Create a system
- Find a rhythm
- Coordinate
- Listen
- Anticipate what's ahead
- Preparedness
- Timing
- Patience
- Consistency
- Creativity
- Adjust the goal
- Alter the approach/method
- Have FUN!
- Respect others by "passing off kindly"
- Be aware of the needs of people around you

Outcomes/Reflections

Participants seem to gain empathy for their peers and an increased understanding of the individual perspectives, roles, and responsibilities that make up their group. Talking about coping skills, stress reduction, and shared responsibilities promotes the health and wellness of the group.

Resources/References: This activity is a variation of a widely used team-building activity that originated with the New Games Foundation (Fluegelman, 1981). This variation with the reflective list and drawing was developed during my time as a recreational therapist at the University of Utah Neuropsychiatric Institute and as a consultant to Middleton Cross Plains Area School District (2002 Middleton High School Health Curriculum). It is also described in *A Teachable Moment* by Cain et al.

Time Management "Full Plate" Lesson

Purpose/Focus: social-emotional learning, 21st century skills, time management, goal-setting, prioritizing goals, coping, executive function, self-regulation, stress management

Materials: paper plates and markers, Time Management worksheets (page 181)

Facilitation Suggestions

- As an entry task, give participants 5 minutes to write and reflect about how they spend their time, using the Time Management worksheet.
- Next involve the group in a short interactive activity/energizer to plant seeds for conversation:
    - Invite students to walk around the room and greet each other.
    - Ask them to focus on one spot in the room while they walk and greet.
    - Next ask them to pat their head as they walk and greet.
    - Then add humming a tune to the tasks.
    - Finally, give them a math problem to figure out in their head as they walk around the room, meet and greet, focus on one spot, pat their head, and hum.
- Ask group members to stop and have them form a circle and share one word to describe the experience. This usually leads to some interesting discussions on multi-tasking—the stress of having multiple responsibilities.
- Introduce the analogy of "having a full plate" to represent one's responsibilities and commitments. Ask participants to think about what is on their plate.
- Hand out paper plates and markers. Ask students to decorate the paper plates with symbols/words that represent the obligations, tasks, and interests that are on their plate.
- Have group members circle up and share some (or at least one) of the items that are on their plates. They will find many commonalities and, at the same time, learn about some of the unique stressors, responsibilities, and interests of their peers. This opens a discussion on how to handle responsibilities and how the group can support each other.
- The time log on the worksheet can be used for a few days as a reflection on how they spend their time.

Resources/References: This activity and reflective prompts were developed with input from teachers and students at Twin Valley Middle School in Whitingham, Vermont. Jennifer Lara developed the plate metaphor (*A Teachable Moment* by Cain et al.), and the students suggested using real paper plates.

Time Management

What do you HAVE to get done each day after school?

Do you use your time wisely? How do you spend time in advisory/study, homeroom, transit to/from school or sports?

Do you have a quiet place to do homework?

What assignments do you put off? Do you work on your favorite subject first?

Time Log

Time	Activity/Task Description	Amount of Time Spent	Importance	Is It Necessary to Success?	Is It a Commitment I Have to Keep?

How Long is a Minute? Can you Feel a Minute?

Purpose/Focus: time management, writing or discussion prompt, perceptions vs. reality, construct of time, self-regulation, reflection, quick energizer

Materials: Stopwatch or other timer

This activity jumpstarts reflection and conversation about perception of time and how our perceptions differ from others. It could be a part of your discussions on time management or used as part of a science lesson on units of time. Science teachers might use this to introduce the idea that the constructs of hour, minute, and second are a human invention. Nothing in nature indicates a 60-minute hour with minutes made up of 60 seconds. The concept of a 24-hour day was developed by the Ancient Egyptians and has been continued based on tradition. It wasn't until modern day inventions such as mechanical clocks and then atomic clocks that the uniform standard for measuring time was developed.

Facilitation Suggestions

- Frame the group conversation to fit your subject matter.
- Ask the participants if they think they know how long a minute lasts, or think they can feel how long a minute lasts.
- Next, have them stand up (or raise their hands). Let them know that, when you give the signal, they are to individually decide when they think a minute has passed and indicate this by sitting down (or lowering their hand).

Outcomes/Reflections

Usually, there is quite a variation in perception of time among group members. Often participants are surprised when the timer lets them know a minute has really passed. Ask them to talk about what they felt as they waited. Possible questions include: What did they observe on the part of their peers? Why does the agreement on these increments of time help us in life? How does it get in our way? What role does time play in your life? How do you manage time? Are you someone who is usually late or early to appointments? Can you relax and do nothing when you wait? How does being kept waiting make you feel?

Variation

Have group members write for one minute, then move around for one minute and reflect on whether their perceptions changed.

Resources/References: I learned this at a conference from Ezra Holland. The Chicago Museum of Science and Industry has a science and technology variation on their website: www.mischicago.org

Headlines: Newspaper Article 10 Years Out

Purpose/Focus: goal setting, values clarification, sense of purpose, reflection, mindset, planning, creativity, communication, vision setting

Materials: copies of masthead of local paper, markers, pens/pencils

Participants create a newspaper headline dated ten years in the future that shares something positive they have accomplished and information about their future life. I use this activity with middle school, high school, and adult groups to initiate reflection and discussion on goal setting and identifying and celebrating personal strengths.

Facilitation Suggestions

- Photocopy the masthead of a local paper onto a blank page and change the date to ten years in the future. Copy these blank masthead sheets for students.
- Ask students to imagine themselves ten years from today. What will they be doing? What is their occupation? Where are they living? What is their family life like? What are their hobbies?
- Ask them to be creative and imagine a positive newsworthy story that might involve them ten years from now. This could range from "Local woman saves children from burning building" to "Local man opens new restaurant in town."
- Have them use the space on the reflective writing "masthead" sheet to write this new story about themselves. Ask them to include information about their personal life such as where they live, their occupation, hobbies, etc.
- The creative writing and illustration parts of this activity can take at least 45 minutes to an hour. So make sure to allot enough time. I often spread this between two class periods, or use a block period.
- After students finish their article, have them share it with their peers either as a whole class or in small groups.

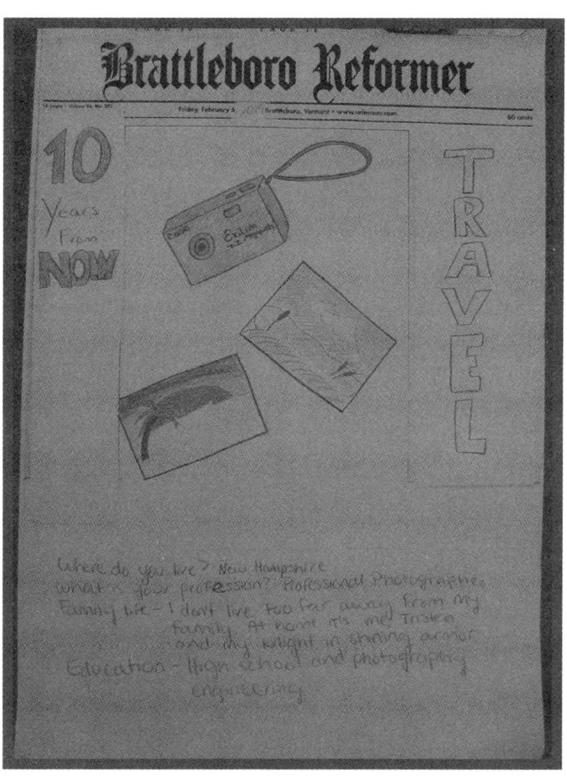

Outcomes/Reflections

Over the years, I continue to be impressed with the creativity and attention students give this activity. In most cases they are very eager to share with their peers.

Resources/References: I used this activity working with the YWCA's "Peer Approach to Counseling With Teens" program and as recreational therapist in mental health treatment. Origin unknown.

Just for Fun With Some Learning on the Side

Large Group Games for Field Day, After School, Family Night, and Recess

I am often asked to facilitate whole-school events and "step-up days" for elementary students transitioning to middle school and middle-schoolers to high school. Most of the games in this book could be appropriate for any of these settings—especially Rock-Paper-Scissors Tag, Have You Ever, and Tin Can Pass. Here are a few other favorites that require more open space, movement, and pure play than many of the classroom focused activities.

Alaskan Baseball (or Chuck the Chicken)

Purpose/Focus: playful learning, social-emotional learning, fair play, executive function, following directions, taking turns, high physical activity level/gymnasium/field game, community building

Material: a rubber chicken or other interesting throwable object

There is something irresistible about the playfulness of this game and its equalizing ground rules. No one needs to be an athlete to be an active participant, so people of all ages and abilities enjoy the game—and it can easily be adapted. I like using it with multi-age groups and having older students help younger ones learn rules of play, cooperation, and turn taking.

Facilitation Suggestions

- Start with two teams.
- Basically a participant from one team throws an object (preferably a humorous one like a rubber chicken), and the other team retrieves it.
- The throwing team forms a circle next to the thrower. After the throw, every time he/she completely runs around the circle it counts as a run.
- Meanwhile, the other team chases the object and forms a line behind the person that gets to the object first. They then alternate passing the object between their legs and over their heads until it reaches the last person, where it is then passed overhead back to the first person in line.

- When the first person gets the object, the team yells out, and the other team stops counting runs.
- The first person in line who ended up with the object now has his team form a circle next to him. He then throws the object, and the process reverses.

Outcomes/Reflections
Though most of the time this is about pure fun, physical activity, and collaboration, it can also be an opportunity to practice conflict-resolutions skills around turn-taking and rules of play.

Resources/References: This activity is very popular in the camp and adventure education field.

Invisible Clay

Purpose/Focus: creativity/imagination, communication, playful learning, collaboration, focus, 21st century skills innovation, turn taking, transitions, peer interaction, imagery, community building

This playful activity is great when you have a few moments left in class or advisory group, or as an introduction to a conversation on innovation, creativity, improvisation.

Facilitation Suggestions
- Ask the group to form a circle.
- Explain that you all are going to play a game with an invisible ball of clay—demonstrate that are you holding this ball of clay.
- Inform the group that the clay, when it is passed, can increase or decrease in size depending on the whim of each sculptor (the person about to pass the clay).

- The leader tosses their piece of clay to another player who catches in a way to show what size the clay is. They might grab it with one hand, or struggle under the "weight" of the clay. If a player doesn't want to sculpt they can quickly pass it to someone else.

Resources/References: I learned this activity from a participant in a workshop many years ago.

Elbow Tag

Purpose/Focus: playful learning, high activity/field/gymnasium game, community building, social–emotional learning, executive function, appropriate touch, boundaries, turn taking, responsibility

Materials: boundary markers to delineate the playing area

This high-energy tag game is perfect for field days, physical education classes, recess, and even adult team-building. I like playing with multi-age groups so that older group members can help younger learners understand the rules of play and model positive sportspersonship.

Facilitation Suggestions

- Create a playing area with barriers, keeping in mind how much physical activity is needed.
- Choose one person to be the tagger and one person to be the runner (walker).
- The tagger chases the runner. When the runner is tagged, they exchange roles.
- Have everyone else divide into pairs. Have the pairs lock elbows and stand in one spot throughout the playing area.
- The runner is safe when they lock elbows with one of the other pairs on the field.
- When they lock elbows with another pair to become "safe," that "knocks" the partner on the other side of the pair off. They now become the runner.

- If one person is the tagger too long, call "reverse," and the tagger and runner change roles.

Variations

For an easier variation, the partners can stand in a circle. For a more advanced challenge, have the elbow-locked pairs walk around while the tagger and runner weave in and out.

Outcomes/Reflections

This game is perfect for older elementary students who are starting to form cliques—participants stop worrying about whether they are paired up with a boy or girl or their friend.

Resources/References: This classic tag game is described in *More New Games* from the New Games Foundation, 1981, and it is also written up in Rohnke's *The Bottomless Bag Again 2E,* 1994.

Triangle Tag

Purpose/Focus: playful learning, cooperation, collaboration, fair play, executive function, self-regulation, peer interactions, turn taking, appropriate touch/boundaries

I use this activity when I want a higher level of physical activity but have limited space. It works with small groups or when a large group is divided into smaller groups. I use it most often with elementary and middle school groups for after-school programs and field days. It's good for youth of multiple ages who can learn from each other. It can also be fun with adults.

Facilitation Suggestions

- Divide participants into groups of four.
- Have three participants in each group hold hands in a triangle facing each other.
- The fourth person is the chaser and remains outside the triangle.
- Designate one person of the triangle as the target that the chaser will try to tag.
- The triangle holding hands will work together to try to protect the taggee.
- When the target is tagged, group members switch positions.

Outcomes/Reflections

With young children the focus often is around learning to take turns and self-regulation. It is a great opportunity to practice helping each other, and learning appropriate boundaries.

Resources/References: Adapted from *New Games* (Fleugelman).

Games for Transitions/Fun Filler Activities

Take advantage of transition times and the extra 5 or 10 minutes that come up unexpectedly during the day with a community-building activity. Some of my favorite community-building activities are perfect to use when you finish a project or lesson early, as students are coming into advisory, or as after school program attendees are coming into the program or finishing snacks. Many are self-directed—once taught, groups can run the activity on their own and "pick it up" the next day. These are either old classics from the camping field or can be traced back to the New Games Foundation. They are popular with the high school students that I mentor, they seem to enjoy them as much as their young charges.

Mystery Leader

Purpose/Focus: playful learning, transitions, social-emotional learning, executive function, self-regulation, following directions, leadership/followership, focus, taking turns, collaboration

This classic camp game works well in a variety of settings with all age groups. High-schoolers love to play this as much as kindergartners. It's a great transitional game—perfect when there is just a short time left in class or advisory group. And, groups can play it on their own.

Facilitation Suggestions

- Have the group form a circle. Ask for a volunteer to become the guesser—the guesser closes his/her eyes or steps out of the room while the group chooses a mystery leader.

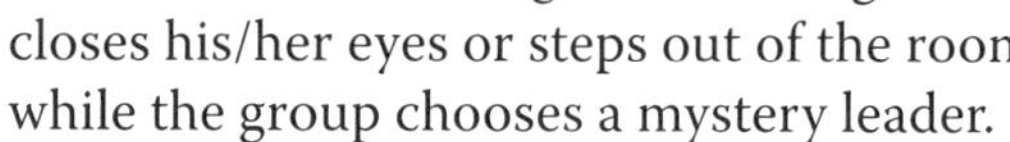

- The mystery leader will subtly lead a series of motions (e.g., snapping fingers, patting their belly, slapping knees), which everyone in the circle will imitate.
- The guesser has three guesses to try to figure out who the leader is.
- The leader tries to change the actions without being detected.

Resources/References: This is known by various names in the recreation and camping field.

Who Has It?

Purpose/Focus: playful learning, community building, social-emotional learning, executive function, self-regulation, turn taking, fair play, collaboration

Materials: any small object that someone might have in their pocket (i.e., quarter, stone, marble)

This classic from the New Games Foundation can be done in any space with a small object from your pocket or nature. Because it involves intrigue, it is a favorite with all ages.

Facilitation Suggestions

- The "It" person stands in the middle of a circle of players.
- While It closes his or her eyes, the rest of the group passes the small object from person to person around the circle.
- After a few moments, group members let It know to open his or her eyes.
- As the group continues to pass the object, It gets three guesses to figure out who has it.

Smaug's Jewels

Purpose/Focus: playful learning, social-emotional learning, executive function, self-regulation, turn taking, collaboration, fair play, appropriate touch

Materials: a bandana, stuffed animal or other soft easy-to-grasp "treasure"

I smile when I look back at this game that appeared in the *New Games* (1981) book when Tolkien's fantasy classics came out in paperback and *The Hobbit* (1977) cartoon movie was released. Here we are full circle, more than 30 years later, with *The Hobbit* prominent in popular culture again, making it a perfect time to play Smaug's Jewels.

Facilitation Suggestions

- One person is chosen to play Smaug the Dragon who will guard the "treasure."
- The rest of the group forms a circle around Smaug, who stands guard over his/her treasure.
- The group members in the circle try to steal the treasure without being tagged by Smaug.
- If a participant is tagged by Smaug, they are instantly frozen into place—but not for long because the game usually is over very quickly.
- The person who steals the treasure gets to be Smaug.

Outcomes/Reflections

This is a great activity for exploring rules of play, self-regulation, social-emotional skills and safety with younger children. Once I set up parameters, I step back and let them adjust the rules as they need, and I encourage them to mediate conflicts over who was tagged for themselves.

King Frog/Queen Elephant/Animal Kingdom Game

Purpose/Focus: playful learning, movement, community building, focus, communication, transitions, executive function, fair play, quick energizer

This is a great "filler" activity because once the parameters are established, it can be picked up at any time and last a long time with extra challenges added as the group progresses. There are many variations of this social memory game in the camp and recreation field. This is the simple version I use with groups for transition times.

Facilitation Suggestions

- The game begins with everyone seated in a circle.
- Each person selects an animal to imitate with a sign or simple physical gesture.
- Each person shares his/her sign with the group, and group members try their best to remember other player's signs.
- One player is the King Frog, whose symbol is to clap hands and then move the top hand away as if jumping off a lily pad.
- King Frog starts off with the frog sign immediately followed by someone else's sign. The player whose sign was made must immediately make his/her sign followed immediately by another player's sign. Play continues in this fashion.
- If any player hesitates and/or misses his cue, that player proceeds to the end of the circle, farthest away from King Frog.
- Everyone else moves over one spot and takes the animal sign attached to that seat.
- The goal is to become the King Frog.

Outcomes/Reflections

I start simply, just getting the animals set up and letting the group practice awhile before adding variations. Like all collaborative games, the process is more important than the outcome (e.g., laughter, use of memory, positive social interaction, and maybe some peer helping).

Silly Veggies Variation

- In this variation, everyone comes up with his/her own vegetable.
- The first person starts by saying their vegetable name twice and another person's vegetable name twice—with their lips over their teeth.
- Having their lips over their teeth makes it very hard not to laugh.

Resources/References: This game has been adapted over the years by camp counselors and educators in various settings.

Coming and Going of the Rain

Purpose/Focus: active engagement, movement, collaboration, active listening, focus, social-emotional learning, taking turns, communication, self-regulation, quick energizer

This rhythm and sequencing game is another classic from the New Games Foundation. I like to use this with early elementary students to practice executive functioning—following directions, waiting for a cue, receptive communication and turn taking—but groups of all ages can enjoy the process of making rain!

Facilitation Suggestions

- Have participants sit in a circle.
- Demonstrate the different movements and sounds involved in making rain: rubbing palms together, snapping fingers, hand clapping, thigh slapping and foot stomping.
- Instruct the group that everyone will make rain by repeating the sound/movements of the person leading the activity.
- Ask everyone to close their eyes and pause for a few moments of quiet.
- Then the rainstorm begins as the leader rubs his or her palms together, back and forth.
- The person to his or her left joins him and then the person to his or her left and so on around the circle till everyone is rubbing palms and the drizzling rain builds in intensity.
- When it has made the full circle, the leader starts snapping their fingers. One by one around the circle, group members replace palm rubbing with finger snapping and the rain turns into a steady patter.
- Next the leader switches to hand clapping and then thigh slapping to intensify the storm.
- Finally the thunder crashes as the group stomps their feet.
- The storm then subsides in reverse, from foot stomping to thigh slapping to hand clapping, finger snapping and palm rubbing.

Outcomes/Reflections

This can be a really nice closing or calming activity. With young children it is a challenge to wait for the cues, but great practice! I often introduce this along with the sports fan wave as a way to teach waiting and following cues.

Run and Scream

Purpose/Focus: high physical activity level/ gymnasium or field game, quick energizer, transitions

This is playful "filler" activity and great for expending energy after a day in the classroom. I use it with very young children, adolescents, AND adults.

Facilitation Suggestions

- Have the group line up, take a deep breath, and then run and scream at the same time.
- Whoever can go furthest on one-breath wins!

Resources/References: I learned this from Karl Rohnke who calls it Run, Shout, Knock-Yourself-Out.

Gotcha

Purpose/Focus: playful learning, community building, appropriate touch, quick energizer, focus, following directions, trust, closing, transitional, collaboration

Facilitation Suggestions

- Have the group form a circle. Ask everyone to put their right hand palm up toward the person on their right and their left index finger pointing down over the hand of the person to their left.

- When the facilitator says, "Gotcha," group members try to grab the finger of the person next to them without getting their own finger caught. Then switch hands.
- In chapter three, I shared this as a partner activity during concentric circles. It works nicely as a quick group closing to bring everyone together and get them laughing.

Resource: Variations of this favorite ice-breaker are described in *Funn N' Games* and *A Small Book about Large Group Games.*

Summary

Brain researchers tell us that executive functions can be learned and improved throughout life with practice. These activities are excellent tools to help learners practice and improve the essential social and emotional skills that will help them thrive in 21st century society. They can be blended into faculty community-building, parent nights, college student life and many other team-building events. Educators can use these activities at the start of the school year to build a strong collaborative learning community, or better yet, blend them throughout so learners can practice and develop these important social-emotional skills over time. Reflection turns this practice into long-term memories, skills acquisitions, and habits that transfer to real life and future learning. In chapter nine, we explore reflective activities to help learners make connections from the skills they use in the games we presented here to real life situations.

chapter 9

Bringing Learning to Life With Reflection

Creating Meaningful, Lasting Lessons With Reflective Practice

Snapshot: The school faculty is gathered for a staff meeting. They are sitting in small groups spread around the room with sets of postcards in front of them. The teachers are taking turns holding up a card chosen a few moments ago to represent a strength they bring to their practice as an educator. Their colleagues are focused on guessing why they chose the card, based on what they know about their co-worker. These faculty members, who have been through a lot of stress in the past year with leadership changes in their school, are enjoying the opportunity to give and receive feedback that honors the strengths and contributions of their teaching team.

The Key Ingredient

Reflection is called by many different names: processing, reviewing, and debriefing to name a few. As discussed in chapter four, I moved toward referring to this key ingredient in teaching and group facilitation as "reflection" or "reflective practice." These terms imply an ongoing, meaningful practice that is woven throughout learning experiences. Reflection is defined as deliberately focusing on thinking about what one has been doing—the intentional attempt to synthesize, abstract, and articulate key lessons taught by experience (Di Stefano et al., 2014).

> We don't learn from experience. We learn from reflecting on experience. —**attributed to John Dewey**

In my view, reflection is the key ingredient in teaching and learning. A century ago, John Dewey emphasized the importance of engaging learners in reflection in order to create meaningful connections between the lesson or learning experience and real life and future learning (1916, 1925). Reflection is at the heart of experiential education philosophy. This enduring belief about reflection as a key component to learning now has scientific evidence to support it.

As this book is being completed, researchers from Harvard Business School, The University of North Carolina, and HEC Paris have released an exciting new study about the impact of reflection on learning outcomes (Di Stefano et al., 2014). Their study is one of the first, if not the first, empirical test on the effect of reflective practice on performance, exploring the idea that learning is enhanced by intentional reflection. This study which combined laboratory and field experiments found a significant improvement in performance when learning-by-doing was coupled with reflection. These researchers hypothesize that the process of reflection builds one's self-efficacy or confidence in the ability to achieve a goal. They argue that their research demonstrates that reflection plays a key role in enhancing learning outcomes. Learning generated by reflection, coupled with experience, will lead to greater improvement in problem-solving capacity as compared to learning generated by experience alone. Individuals perform significantly better on subsequent tasks when they think about what they learned from the task they completed. The researchers state "learning generated by 'doing' becomes more effective if deliberately coupled with the controlled, conscious attempt at learning by 'thinking'" (p. 4).

This research is important because it gives empirical evidence to support the enduring beliefs put forward by educational innovators such as John Dewey (1910, 1934) and Donald Schon (1983) that reflection is integral to learning. Secondly, their research on the connection between reflection and self-efficacy helps educators understand the value of reflection and the role self-efficacy plays.

Educational neuroscientists tell us that intentional engagement in reflective practice is key to "cementing" learning. They emphasize the importance of creating multiple pathways to learning, facilitating "patterning," and providing ongoing feedback throughout educational experiences to promote strengthening of neural networks. When learners can see their progression through learning and receive ongoing feedback about what they did well and what they need to do to improve, intrinsic motivation is increased because they see the work as an "achievable challenge" (Willis, 2011a).

The metacognition or "thinking about thinking" involved in reflection helps people become aware of their learning capacities. Educator and author Sam Wang points out that, when we understand that intelligence is not necessarily a fixed quality, we learn more. In a recent conversation Dr. Wang stated, "To me this is an impressive meta-result. A possible explanation is that when students are made explicitly aware of the fact that active experiences change the brain, it opens them to the possibility of doing so" (Wang, personal communication, April 19th, 2014).

Reflection can also serve the practical purpose of providing important formative or summative assessment data for educators. When engaged in meaningful reflection, both participants and educators learn about group and individual progress as well as areas that need more focus and attention. Reflective activities involving writing or artwork can be a record of lessons learned that can be referred back to by participants or educators for future learning.

Our brains search for meaning and relevancy in lessons. The process of reflection helps learners understand HOW information is relevant to them and specific ways they will apply the information learned in the future. Reflective practice fosters an emotional connection to the material being learned. It helps people internalize the lesson, creating ownership over learning. This enhances the ability for participants to grow and change through their experiences and develop **insight**, one of the most important lifelong skills to acquire. When we engage learners in ongoing reflection in our programs and courses, we help them become more reflective, introspective learners in the future. The goal of reflective practice is to help participants recognize their strengths, manage their behaviors, and learn to apply skills and insights learned in one situation to the next. This chapter is devoted to a variety of tools and techniques to facilitate meaningful ongoing reflection and help learners develop the life-long practice of reflection. Meaningful reflection can and should be woven throughout the entire experience using a variety of methods. As you see in this book, there is a great deal of cross-referencing between this chapter on reflection and chapter five, which focused on strong beginnings.

The Challenges of Facilitating Reflection

Educators recognize the value of reflection and processing, but they find facilitating it is one of the most challenging aspects of teaching, counseling, and group work. Many report that it sometimes feels like a chore or like pulling teeth—they receive blank stares, the same group members always speak, or the answers are what learners think the facilitator or teacher wants to hear. Others share that they run out of time for reflection despite their best intentions.

I believe these difficulties arise from a "one tool" approach to reflection. Many of us were taught in our early training how important reflection is, but only taught one method for facilitating it—the traditional, didactic, verbal Q&A session led by the facilitator with directive questions. Some of us were given Terry Borton's helpful model of structuring questions which encouraged educators to design questions that cover "what happened," "so what," and "now what" (1970). The aim of his approach was to help facilitator's lead more meaningful reflection by covering the emotional aspects and application of experiences, including the story of what happened in an experience, the emotional impact on the individuals, and finally the key aspect of application to real life. We were told that if we just asked open-ended questions that covered these questions, we would effectively lead meaningful reflective dialogue.

I and many of my peers found that, in real life, a verbally-focused, facilitator-led question-and-answer is not always the most effective technique for eliciting meaningful, reflective dialogue especially when learners are new to the group. Putting learners on the spot with directive,

> There's all the difference in the world between having something to say, and having to say something."
>
> —**John Dewey**

verbal questions in a group-sharing circle can inhibit meaningful reflection. Though there is obviously a time and place for a didactic question and answer session, it is not always best for engaging learners in reflective group dialogue.

The other weakness of this model is that the facilitator/teacher may design questions around what they think the group learned or was supposed to learn. Following this planned line of questioning, the discussion might miss the important moments and lessons the group actually experienced. Often when facilitators/teachers use this method they find themselves interpreting the experience for group members instead of eliciting meaningful reflection.

After experiencing this myself in my own groups, I started experimenting with alternative approaches including metaphoric objects, active partner sharing strategies, artwork, and repurposing icebreakers and games as interactive dialogue, review and reflection methods to facilitate this process. Using these varied techniques, the reflective dialogue became much richer, the transfer to real life became clearer, and there was more participation by all group members. Best of all, there was more ownership and buy-in on the part of learners. As years go by, I gravitate toward more participant-directed approaches to facilitating reflection.

Student or Participant-Directed Reflection

Meaningful participant-centered reflection is initiated or guided by the facilitator/teacher but not led or directed by the facilitator. Learners are engaged in ways that they can take ownership of the interpretation and application of learning rather than the facilitator possibly inserting his or her own analysis or agenda. Steven Simpson promotes this approach in his book, *The Leader Who is Hardly Known* (2003), suggesting that facilitators consciously use methods that put more control in the learners' hands and give them the tools for interpreting their experiences/feelings/goals. Reflection can be intentionally incorporated into the icebreakers we facilitate, it can happen in the middle of lessons or group initiatives as well as after the lesson or activity. With this type of ongoing practice, group members begin to initiate reflective discussions on their own and the facilitator can step into the background. Most importantly, this approach

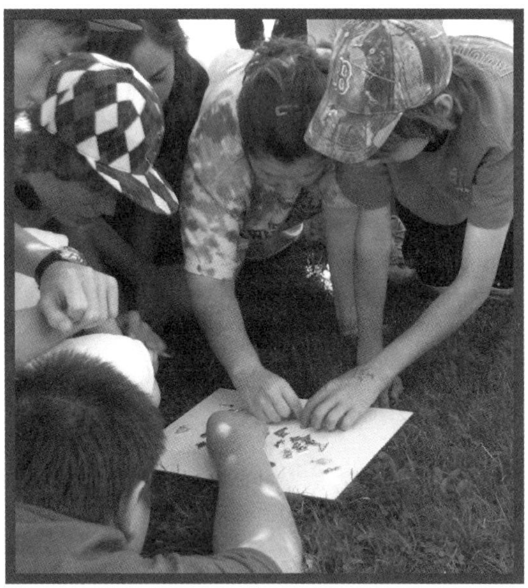

helps participants continue reflective practice long after the experience is over. This chapter explores different methods for facilitating this key ingredient in education and group work.

Environment and Sequencing

The importance of building a positive and supportive environment and sequencing and scaffolding learning experiences is especially true in facilitating reflection. Learners need to be in a safe and supportive environment in order to constructively share their feelings. In our society, we are bombarded with information and sensory stimulation and generally take little time for quiet reflection. Reflection is a practice. We build comfort with this practice by introducing it to our groups incrementally, starting with simple object choices and partner sharing before asking them to share with a large group.

Make sure participants know what is going to be done with the information shared and that they have choices around how much and when they want to share. Follow up or respond to difficult moments or when something hurtful is said and foster positive group norms within a group to ensure quality sharing. By varying methods and including active reflection, partner pair and share, writing, and dialogue, we are more likely to engage learners and help them view reflection as an exciting and dynamic part of group experience.

Weaving Reflection Throughout

Reflection is often left to the end of an experience, lesson or activity. Often it is overlooked or shortened because of logistical changes or because it feels like a follow-up chore. About ten years ago, Luk Peeters, a therapist and experiential educator from Belgium, shared with me that he believes that the optimal place to capture a lesson is WHILE it is happening. He later shared this statement in an interview for my Inspired Educator Blog:

> *In order for change to happen the optimal time is whilst experiencing and at that time being able to open up for the new or the not-alike. Afterwards we often 'think' we have learned something but in the next situation we do exactly as before, and that is because the actual emotional schemes that direct our behavior have not changed. It is like trying to change a software program in a computer. That is only possible if the computer is up and running.* —Luk Peeters, 2010

His comments sparked my thinking about how to take advantage of those opportunities. I had occasionally been stopping groups in the midst of an activity to discuss and reflect, but usually only when things weren't going well. Why not stop a group midway when things are going beautifully so they can remember what if felt like, looked like, and sounded like?

I started trying to weave in some brief "stop actions" in the midst of a challenge for groups to stop, look around, and identify what is working. This was very effective for a group that had struggled at the beginning and was now moving forward to identify what it was that had changed within their group process. It seemed that stopping mid-activity allowed group members to take mental snapshots of key teachable moments and better apply them in subsequent activities.

Metaphoric image cards and objects are effective ways to process "midstream." Lay out cards or objects at a transition point in the middle of a program and ask groups to come to agreement on three cards that represent three important skills they used in the activity that they think will be useful in upcoming activities. Ask them to carry these cards or objects with them to the next activity to keep the group from falling into old habits. Find ways to integrate them into the next activity or refer back to them to illustrate transfer of skills/learning to future situations.

Other opportunities for weaving reflection into the activity include using objects in frontloading an activity. For example, have group members choose objects that represent goals for their program, workplace, or other situation and then incorporate them into the challenge itself in some way. There are many creative ways to integrate reflective practice into the action phase of learning. Regardless of how you do it, I think you will find that groups engage in more meaningful reflective discussions spontaneously throughout their learning experiences if you mix up your approaches to facilitating reflection.

Reflective Activities by Type

You will notice that most of the activities discussed in this chapter have been offered in previous chapters for other purposes as well as reflection. This was intentional, to illustrate that reflection can and should be an ongoing integrative part of a programs, not just something left for the end as a follow up. This chapter highlights how these activities and techniques can be used when reflection is the primary purpose. With multiple pathways to learning in mind, this chapter categorizes techniques by "type" to help vary strategies. In many cases, you will find a reference to the detailed activity descriptor in a different chapter. In the following pages, you will find reflection variations and extensions as well as real life examples.

- Reflective Prompts and Context Setting Activities
- Pair Share: Partner and Small Group Reflection Strategies
- Re-Purposed Icebreakers: Games as Reflection Activities
- Twists on the Traditional Sharing Circle and Q & A Sessions
- Creating Meaning Through Metaphor
  - Metaphoric Images and Objects Activities
  - Metaphoric Activities and Games That Represent a Lesson
- Art and Creative Expression in Reflection
- Self-Reflection Techniques
- Mementos of Experience: Carrying the Lesson Forward

Reflective Prompts and Context Setting Activities

These are the activities that provide a context, set the stage, or frame an upcoming or ongoing learning experience. They plant the seed for further conversation and reflection on a subject or initiate thinking about expectations or goals for an upcoming experience. They can even begin an inquiry process into a subject, inspiring learners to start asking questions about the subject or topic at hand (this can also be a formative assessment). Activities used in this way can become an established tool that is referred to throughout a program and then reflected upon again at the end.

I often **bookend** experiences by wrapping up with the same context-setting activity that I used at the beginning. This ties a lesson together, strengthening connections to learning and helping participants see how far they have come. Based on what educational neuroscientists are telling us, this kind of context-setting, reflective activity facilitates relevancy and meaning making, promotes patterning, strengthens multiple pathways to learning, and enhances long-term memory. The following context-setting activities from chapter two are specific examples of using reflective activities with groups.

Reflection Starter Buttons (page 76)

The Conversation-Starter, pin-back buttons activity can be an engaging reflection tool to jumpstart a conversation about roles in a group, progress on a task, or participants' feelings about the group experience. Lately, I have been using the Conversation Starter buttons with middle and high school students as a way to check in on progress around group projects.

Recently, I used the buttons with middle school science students to reflect on their cross-curricular, 8-week-long collaborative bridge-building project. I asked them each to choose a button that represented their role, perspective, or personal experience with the project. It led to a meaningful discussion about the dynamics within small groups and the challenges that arise in working together. Students appropriately gave and received some difficult feedback about each others' behavior within those groups in a positive way. The buttons added some humor to start a difficult conversation, and they provided a reference point to open the discussion.

When I attended their "Bridge Night" presentation a couple weeks later, one group of students shared with me that the reflective button exercise and the solution-oriented conversation that followed led them to make an agreement with each other to work differently throughout the remaining days of the project. They ended up being the group with the first bridge standing and one of the strongest bridges. Most importantly, they were able to articulate the value of what they learned from working through the challenges together.

A student chose the "Why am I the only one working here?" button, stating she was the only person in her group who really worked, and it was because of her efforts their bridge was completed. A boy in her group who rarely speaks in class chose the "One of the little people who makes things happen" button. He responded, "I could have helped, but you wouldn't let me. I don't think you trusted me." This boy received special education help, and it was evident the girl did not view his capabilities as equal to her own. In fact, he was very capable at building, designing, and math challenges. The buttons allowed them to call each other on their behavior in a way they could hear. Months later this girl wrote a postcard to her future self—"The most important lesson I learned in middle school is that other people besides me have something to offer, and that I need to do a better job of listening and being open to the ideas of others—even people who aren't my friends or who I think are different."

Personalities Pin-Back Buttons (see Conversation Starter Buttons on page 76)

Recently I was asked to facilitate a team-building program with a school administrative team. One of their main goals for the session was to reflect on their roles in the group and the parts they were going to play in finishing and sharing the district's strategic plan.

- I gathered "Personality" Pin Back Buttons by Jane Jenni (see resources) with graphics such as "Social Butterfly," "Squeaky Wheel," "Open Book," "Watch Dog," "Work Horse," "Organic," and "Queen of Crisis." Then I displayed them by the coffee as participants came in for the meeting that morning.
- I asked participants to pick one that represented the role they see themselves taking within the team.
- Our introductory, group-building activities centered on sharing these buttons. Throughout the day, as people expressed their thoughts, opinions and rationale for an idea, they often referred back to their button.

Quotes (page 75)

Quotes are an especially effective tool for context setting or themed reflection on a specific topic being explored by a group. For example, I found quotes centered on change, perseverance, and flexibility and printed them on cardstock for a back-to-school in-service for faculty going through consolidation and reconstruction of their school. I used the quotes along with partner sharing activities to spark reflection on coping with change and taking advantage of the opportunities that arise with it. With a guidance group exploring school climate and citizenship, I gathered quotes about empathy, caring and responsibility toward others. I used them as a transition reflection activity in the midst of the day long retreat. We found that the quotes helped students articulate what empathy means. This method allows facilitators to guide the direction of the reflective conversation while still promoting participant choice, control and creativity.

Ideas for Using Quotes in Reflection

- Display quotes and ask participants to choose one that represents the unique perspective or philosophy they bring to the group. Ask them to share with a partner, write about it, or quietly reflect about it.
- Quotes make great writing prompts. Have learners write about their quote as part of a reflective writing exercise.
- Use quotes after an activity, during transition times or as a closing. Ask participants to choose one that represents a key learning or new perspective they will be taking away from the experience.
- Allowing participants to take the quote card as a memento of the experience can serve as a reminder to reinforce lessons or goals.
- To reflect on leadership roles or perspectives on teaching ask group members to choose a quote that represents
  - the approach or role they took in the group experience,
  - a quality they think is most important in a leader or teacher, or
  - their personal philosophy of leadership or teaching.

- Ask participants to find their own quote to bring to the group to share.

Resources: There are some great quote books and quote websites. I especially like to seek out vintage quote books in used bookstores. Enjoy your search!

Toolbox Reflection (page 73)

Over the years, I have experimented with ways to structure and facilitate my annual "Facilitator's Toolbox" workshop at the Association for Challenge Course Technology conference. This workshop is a forum for educators to bring and share their favorite tools and techniques and ask and explore questions of other professionals in the field. I decided to dump all of the items out of the old toolbox I used as a "processing treasure chest" (e.g., slinky, binoculars, mardi gras beads, old camera, keys). I re-filled the box with items that could be defined as "tools," ranging from Allen wrenches, a whisk, ruler, oil can, hammer, glue, magnifying glass, extension cord, etc.

As participants entered the workshop I asked them to pick a tool that represented the most important "tool" to have in their facilitation toolkit. It could represent an activity or strategy, a frame of mind, philosophical approach, etc. The tools set the context and theme for the workshop and brought to the forefront what the group had to offer each other and what they wanted to focus on during our time together.

Computer Keyboard Keys (page 77)

In chapter four, I shared about using this upcycled reflection tool as an introductory or context-setting activity at the beginning of a program. It also makes for an effective reflection activity.

Sandi Lindgren who is a social worker with an organization called I Support YOUth! shared her experience with this activity in my Inspired Educator blog (April, 2011):

> *I used the keyboard keys yesterday and the group loved it! I was speaking to a senior class of social workers at a local college about research/surveys (sharing examples from my dissertation) and how to find your dream job as a social worker. I used the keys in the beginning as an introduction, and asked them to choose a key that best represented where they currently were in this class' research process. The professor was delighted with the on-point honest sharing that occurred. I was then able to reference their examples later on in the session.*
>
> *Examples included: The space key for taking a break and not really doing much with homework. The S key for stress. The tab key for taking a break (the old soft drink tab). The group also later pointed out that the tab key moves one forward. The professor picked the command/apple key, sharing that as a teacher (apple) he was delighted with the learning and processing. The page down key was chosen by someone who felt their process was really slow. The caps lock key was chosen by a student who was so stressed she felt like everyone was talking to them in ALL CAPS—WHICH IS YELLING ON THE SCREEN.*

I had an interesting experience during a facilitation training for corporate executives, students and non-profit organizations. The focus of our day was on expanding their facilitation toolkit. During the lunch break, I sat the keys out and asked group members to pick a key that represented a personal goal they have for improving their group facilitation. One group member chose the F3 key, stating that, on computers, this key opens new search screens. He chose that key to represent his desire to "find the next new way" and be open to experimenting and looking at old problems in a new light. During the remainder of our training, the group used "F3" as code to have an open mind. When someone got stuck in a fixed mindset, another group member said, "Come on, let's F3 it!"

Reflective Writing Prompts (page 79)

As mentioned in chapter four, reflective writing prompts can help prime a group for upcoming conversation, activities, and lessons about a subject. As with other context-setting activities, prompts can be used throughout a program and taken away as a reminder of lessons learned.

Pair Share: Partner and Small Group Reflection Strategies

People gain comfort with reflection if they can start by sharing with just one or two people at a time and work up to conversing in larger groups. We are social beings and brain research shows that we learn and attend more when we are involved in social interaction and movement. A variety of activities commonly used for introductions make great review and reflective dialogue techniques. Everyone has the opportunity to participate, reflect and share without being put on the spot in front of the whole group. These methods work well with large groups and can be used multiple times and woven in at any time during the group's experience together.

These strategies are described in other chapters as introduction or academic review strategies—they are also excellent reflection strategies. Once a group is familiar with them, you can implement them as reflective techniques/conversation starters. I often bookend experiences, using an activity at the beginning of a program or class for introductory dialogue and again at the end of the program as a closing to tie it all together.

Handshake Mingle (page 89)

My favorite ice-breaker/group building and introduction activity is high on my list because of its multi-purpose uses as an icebreaker, as an active academic review technique, and as a partner reflection method. It can be used right after a group problem-solving activity or other learning experience to jumpstart reflective conversation in a playful way or during transitions as a short reflection/check in around a topic. It can also be used as a closing activity to tie the learning all together and help participants leave a program or class period on a positive note, celebrating, reflecting upon and articulating lessons learned and connections to future learning.

Concentric Circles (page 86)

This more reserved active partner dialogue serves the same purpose as Handshake Mingle in a more structured way. The benefit of this method being that participants don't have to find a partner to reflect with as the structure of the activity creates random pairings for partner dialogue. This can be made playful by combining partner activities with the sharing. For more serious conversations, just rotate between the circles. This is an excellent way to pair groups with differing perspectives or experiences with those from the other group by having one group on the inside circle and the other on the outside.

Dominos, Match Game Cards, Playing Cards, Key Matches (page 95)

These methods establish partnerships at the beginning of a program for introductory conversations. At any time a facilitator can ask these partners to find each other for reflective conversations.

Trade and Share (page 91)

This activity doubles as a way to reflect on the strengths or accomplishments of group members and makes a nice closing activity. It is okay for you as the facilitator/leader to not hear all of the reflective conversations in the group. Active pair share methods help people build comfort with sharing one person at time. If something needs to be said to the whole group, this incremental build up to group sharing will allow for that to happen when the time is right.

Re-Purposed Ice-Breaker: Games as Reflection Activities

These active and social activities engage learners in a playful, interactive way. They are, in my experience, one of the best ways to introduce reflective practice incrementally and build a group's comfort with sharing. Participants who don't usually offer insights and answers in a traditional sharing circle will actively share through these kinesthetic activities.

Anyone Who (page 98)

This works well as a reflective or closing activity for large groups to tie it all together, review highlights of the day or lesson, and celebrate program successes. Once a group has engaged in this activity as an icebreaker or for academic review, it is easy to use it again with reflective questions. Groups enjoy this activity enough to want to do it multiple times in different contexts.

- Sometimes I use prepared index cards with reflective questions about the day or program as an option for participants who land on the question-asking spot.
- Participants choose a card or ask a reflective question of their own.
- When a group is comfortable with the setup of the activity, they seem more comfortable with sharing. Ask the person who lands on the question-asking spot to share more about why they moved.
- Some examples of a reflective prompts follow:
  For teacher training: "Anyone who will use a technique they learned in our training today in their classroom." The person who lands on the hot spot has the opportunity to share the activity they will use in their classroom before going on to the next question. This keeps everyone thinking, reflecting and reviewing the highlights of the day.
  For a team-building program: "Anyone who can share a moment when they were proud of this group today." When they land on the spot they share this moment with the group.

Commonalities Mingle (page 97)

As a reflective technique, have participants group by the postcard or object they chose as an entry task or reflection prompt (see page 73) or by their perspectives on an experience or role they took in an experience.

Facilitation Examples

I found that combining the strong beginnings activities shared in chapter four with the Commonalities Mingle is a great large group reflection or context setting activity. The following is an example of how I wove this activity into ongoing, large-group reflection throughout a conference experience.

Facilitation Suggestions:

- At the start of the conference or course, I asked group members as they registered to choose a button that represented their approach or attitude as they begin the day (page 76). Participants organically interacted with others about their buttons throughout the conference, making this a very participant-centered, conversation-starting activity.
- During a lunch break later on in the program, I display Quotables, or quote cards (page 75) and ask group members to choose one that resonated with them.

- At the final keynote or closing event, I laid out postcards and ask group members to choose one they would send to a friend to describe their time at the conference.

Commonalities Mingle can also be used as an interactive and playful closing reflective activity. Have participants group up using reflective commonalities questions/topics such as:

Group up with people who traveled from the same region as you.
Group up with people who chose a similar button as you.
Group up with people who chose a quote for a similar reason as you.
Group up with people who chose a similar postcard as you.

Charades Race as Reflection (page 112)

This game works especially well for reflection at the end of a program, field trip, training, or as a way to celebrate the closing of the school year. Simply ask each group member to write down a highlight of their day, program or experience, or, a moment worth celebrating. Then use these "moments" as the material for the Charades Race.

Facilitation Example: I used Charades Race with a camp staff group that was finishing up a week-long training together. Staff members used content from their training material, camp traditions, and fun moments the group had together during the week. The camp director received valuable feedback on what participants were taking away from the training experience. I continue to use this at the end of workshops or trainings and, most recently, the end of the school year with 8th graders. I ask participants to write a highlight of their time together or an important lesson they are taking away. I then used those terms to play the game. It becomes a playful and fun celebratory and reflective closing activity.

Play Dough Pictionary Reflection (page 110)

This activity can be used in a similar way as Charades Race for reflecting on program experiences. Play dough or clay can be used as a tactile and kinesthetic tool for engaging a group in a variety of creative expression activities to explore the lessons learned from an experience.

Peek-A-Who as Reflection (page 145, 166)

When the blanket drops, the teams describe the accomplishments, and positive contributions of the person on the other team while their teammates guess based on what they experienced with that person during the program.

Twists on the Traditional Sharing Circle and Q & A Sessions

As mentioned previously, traditional Q&A sessions are not right for every situation. However, many facilitators are drawn to this leader-directed method because they want to make sure specific reflective questions/topics are covered, and they want to hear from the entire group at the same time for evaluative purposes. Take the traditional "What?, So What?" and "Now What?" questions and add a little fun with simple props as a novel and participant-centered way to engage the group in reflection. Because these methods add a spontaneity and randomness about who might answer the question, everyone ends up reflecting on the question at hand, not just the hand-raisers. An interesting aspect of these playful methods is that group members who usually chose to pass in formal discussions will willingly share if the spinner or dice points to them. It is as if fate has chosen them to answer, the randomness seems to elicit a sense of choice that is different than a facilitator directly calling upon the participant.

Review and Reflection Processing Dice (page 104)

Elementary and middle school students love to roll the dice, which means they will have more questions to reflect upon. The trick is to roll the question die before rolling the die that decides who is going to answer so that everyone reflects. Steven Simpson, Buzz Bocher and Dan Miller of the Institute for Experiential Education created the *Chiji Processing Dice* that I often use in group facilitation and teaching (see resources). This set of four styrofoam dice include three that have questions covering what happened, so what, and now what. The fourth die has arrows indicating which participant answers. You can easily make dice yourself (recycle small boxes or pieces of Styrofoam) to customize reflection questions or for academic review and reflection. Dry erase dice are available in many educational supply stores that work well for this purpose.

Spinner

Buy arrow spinners from a toy supplier (see resources) and make your own spinner board with different reflection or review questions. The spinner can be passed around the circle or placed in the middle for participants to take turns spinning for their reflective question.

Beach Balls, Soccer Balls, Thumballs

Add reflective prompts to the different panel sections of a beach ball or the pentagons and hexagons of a soccer ball to encourage participation. Answers in Motion LLC created a line of ready-made "Thumballs" for this purpose. These are small, soft soccer-style balls with reflection and conversation prompts around various themes printed on the different sections. When the ball is tossed around the circle, the person receiving it responds to the prompt closest to his or her thumb (see resources). For the beach ball variation see *A Teachable Moment* by Cain et al.

Pass the Knot (page 103)

This simple active conversation activity involves the group quickly passing a rope tied with various knots around the circle. Youth enjoy this activity as opposed to a traditional sharing circle because of the activity, playfulness and random choices of who might answer.

- The teacher or facilitator asks the group a reflective question.
- The group then passes the knot or knots around the circle.
- When the facilitator says "stop," participants with a knot in their hand share a reflection or answer to the specific reflection or review question.

Creating Meaning Through Metaphor

> *I believe that much of the time and attention now given to the preparation and presentation of lessons might be more wisely and profitably expended in training the child's power of imagery and in seeing to it that he was continually forming definite, vivid, and growing images of the various subjects with which he comes in contact in his experience.* —**John Dewey (McDermott, 1981 p. 451)**

Of all of the methods in my teaching and group facilitation toolbox, my collections of metaphoric images and objects are the most powerful for igniting meaningful reflection for both individuals and groups. Reflective conversations sparked by the use of image cards, pictures, artwork, and tangible objects have a richness, depth, substance and level of participation that verbal techniques alone rarely elicit.

Symbols, images and objects capture participants' attention, arouse emotion and help people communicate their thoughts and create meaning from experiences. Humans have long used symbols to understand and communicate abstract concepts that cannot always be translated into words. Participants engage more readily and share more deeply when they have a visual or tactile symbol to represent their thoughts and feelings. Using metaphors and symbols in conversation stimulates multiple areas of the brain involved in other senses (Lacey, Stilla, & Sathian, 2012), making this another effective activity to promote multiple neuronal connections and pathways to learning.

I find there is increased involvement from more introverted or reluctant group members when I use metaphoric object and image activities. When participants can talk about the tan-

gible object or image rather than directly about themselves or their feelings, they seem more comfortable and willing to share with the group. These methods are not only engaging and non-threatening; they allow abundant opportunities for creative, spontaneous, and meaningful interpretation. They promote participant control and ownership in learning experiences. A visual or tactile representation of an experience can be an effective reflective prompt long after the experience is over, reminding a group or participant of the key lesson learned (Stanchfield, 2007).

When groups agree on an object or symbol that represents their collective experience, this symbol can end up carrying a great deal of power for a group. A drawing, sculpture, or collaboratively created symbol can become a kind of mascot representing the group's strengths and achievements, or it can represent a goal to be reached (Cain et al., 2005, p. 42). Younger participants also can benefit from the use of symbols in facilitation as children naturally think in metaphors because they are accustomed to storytelling and using their imaginations.

Although both the image and object activities are similar approaches, I find that groups respond differently to using cards or pictures than they do objects. You can present both object and picture activities in a program and they won't be perceived as a repeated activity; they actually will reach people in different ways (Stanchfield, 2007).

Metaphoric Image and Object Activities

- Pick-A-Postcard, Chiji Cards, Image Cards (page 71)
- Treasure Chest, Miniature Metaphors, Toolbox (page 73)
- Found Objects—objects found in nature, from the room, from participants' book bags Groups can create their own image kits or treasure chests (page 73).
- Keyboard Keys (page 77)

Metaphoric Activities or Games That Represent a Lesson

Many activities act as a metaphor to generate reflection on a specific topic. The following metaphoric activities found in chapter six initiate reflection on communication, community, empathy and balancing priorities. These powerful activities symbolize a life lesson and naturally lend themselves to reflective discussion. The addition of a worksheet or art activity can strengthen the power of the metaphor.

- Telegraph (page 152)
- Community Circles (page 151)
- Communication Breakdown (page 155)
- Keys to Community Building (page 171)
- Full Plate (page 180)
- Helium Hoop (page 162)
- Group Juggle Metaphor Drawing (page 177)

A Success Story from Educator Jennifer Bradford

"During my one-on-one time with a student who has been struggling both academically and socially, I spread out a pile of your Miniature Metaphor trinkets and asked him which of them best represented how he feels about 5th grade. He studied them for awhile and then asked me if he could pick more than one. 'Talk to me,' I said. 'What are you thinking?' He picked three charms: binoculars, a shovel, and a treasure chest.

"His thinking? The binoculars showed that he was looking for something. The treasure chest was what he was looking for. And the shovel was how he would get it. What was he looking for? What did he identify as 'the treasure of 5th grade'?

"Friends. Poignant, yes. Sad, yes. Hopeful sign? Most definitely. The fact that he could articulate what he wanted made it so much more likely that he would achieve his goal. And, for me, the fact that he picked the shovel indicates that he is willing to work for it. And I, for one, believe that he will find it."

Facilitating Suggestions for Individual Reflection

- Invite participants to pick an object or image that best represents
  - their experience,
  - the role they took in the group,
  - something they have achieved,
  - their unique perspective on the group experience,
  - a contribution they made to group process, or
  - their personal strengths
- Participants can draw or write about their object/image in journals or to share with partners in an activity such as Concentric Circles or, if appropriate, with the whole group. Invite them to reflect on a lesson learned or write about how the card or object represents the lesson, a personal goal, or commitment.
- Consider using the postcard to carry learning forward, and ask participants to self-address the card. Have them write what they learned and a commitment to themselves on the card. Then send it to them as a reminder/memento a few months later .
- In some situations, it could be appropriate for group members to choose a symbol/object for another person in the group as a memento or gift representing the strengths or contributions that person lent to the experience.
- If possible, consider allowing participants to keep the object or picture that represents their strength or achievement as a memento/reminder.

Facilitation Suggestions for Group Reflection

- Ask the group to come to agreement on one card/object that represents what the group has achieved together or the strengths or shared vision of the group.
- Have the group pick three objects or images that represent three important skills they used in the activity that they think will be useful in upcoming activities. Invite them to carry these objects with them to the next activity and find a way to integrate or refer back to them to illustrate transfer of skills/learning to future situations. Blending opportunities for reflection on what is happening while it is happening can be very powerful.
- Consider integrating objectives/images into an activity to represent goals or obstacles, or as a tool to tell the story of the group's journey through the activity.
- Group members can use cards/objects to relate the "story" of their experience by lining them up to demonstrate how they progressed through the activity, the day, or the course as a whole.
- My colleague, Dave Lockett, suggests asking groups to choose three cards/objects to tell the story of the their journey. Participants pick cards to describe where the group was when they started, where they have gone or what they have achieved so far, and where they hope to go next as a group (Stanchfield, 2001; Cain et al., 2005). I use this as a check-in halfway through a program or as a closing activity.

Facilitation Suggestions for Group Goal Setting or Group Norms Discussions

- Ask group members to work together to choose a card or object that best represents
  - what they want to achieve together,
  - a symbol of the group's strengths and purpose,
  - the specific group norms, ideal behaviors or characteristics they would like to ensure are present in their group, or
  - the shared vision/goals of the group.

Art and Creative Expression in Reflection

When a symbol is used or created for reflection it leaves the opportunities for meaningful interpretation of an experience wide open, facilitates multiple pathways to learning, and enhances emotional connection, retention and application to real life. Art and creative expression activities are participant-centered, and the learners end up with a tangible reminder of the learning.

Artistic reflection activities can involve a range of mediums and methods—drawing, sculpture, performance art, photography, music, or documentary—the possibilities are endless. As long as the emphasis is taken off artistic skill, most people will readily engage in this kind of

creative activity regardless of their perceptions about their artistic ability. Enhance the process by providing accessible tools and a variety of fun materials and giving short periods of time in a casual atmosphere (Stanchfield, 2001, 2007).

Group drawing can be part of a formal journaling assignment or a playful group activity. Have participants use any or all of a variety of mediums (paint, crayon, markers, chalk, etc.) to create a picture, cartoon, map, or another expression of their experience. My favorite version of this is to have the group (or small teams within a larger group) create the picture together. As a closing activity, I often have small groups/teams quickly work together with fat markers to create a pictorial representation or symbol to represent their day together. The fast pace and large markers keep participants from getting too caught up in artistic ability. There is a lot of humor along with interesting insights when groups share their drawing. In classroom settings, this can be a way to check for understanding and gauge reactions regarding a specific lesson or experience (Cain et al., 2005).

In schools and other longer-term programs, groups can **create a collage or mural** of experience, using a combination of quotes and photographs from their experiences (pictures and quotes cut from magazines and other printed media could be an option). See the Graffiti Wall activity on page 83.

Invite the group to **create a symbol** that represents its experience and achievements. Again, supply a variety of materials (or let them come up with their own) for this project.

Community puzzle pieces (see page 173) can be decorated with representations of each person's strengths and fit together to create a quilt-like mural of the group's strengths and diversity.

Scrapbooking and bookmaking incorporate a variety of artistic methods. A student in one of my programs used a combination of quotes from fellow students, photographs of their experience, drawings from group members, and found objects to create a beautiful scrapbook depicting the group's experience and growth. *The Bookmaking Experience* (Bickle et al., 2012) is an excellent resource for educators looking to engage learners in bookmaking (see resources).

There are numerous ways to involve students in **creative writing**. Groups can create a poem, song, or story about their experience. The middle-schoolers with which I work are adept at creating poems and raps about our field trips or school events. Students could act as reporters for a newspaper and interview fellow participants.

At the end of August, 2011, our Vermont community was hit hard by flooding during Tropical Storm Irene. Homes and businesses were lost; power, phone and Internet were out for many days; and bridges and roads are still being rebuilt. School started a week late, and it seemed important to reflect on the flood and the delayed start along with the normal excitement and stress of the beginning school year. I started our first session with the **Graffiti Wall.**

Students were invited to decorate the wall (butcher paper) with graffiti to describe their experiences, reflections, questions regarding the flood, what it has been like starting school amidst the recovery, and what it felt like to be an 8th grader. Like a real wall of graffiti, others added to or commented on the artwork. The activity initiated meaningful reflective discussion about experiencing the loss, but also witnessed the recovery. One student who lived downtown shared that she was keeping a list of all the items that floated by her house. Students gained an increased sense of compassion and understanding through the flood experience. EVERY student shared an example of service they had offered the community during the crisis, including baby sitting at the Red Cross center so that parents could help clean out homes and businesses, working at the food bank, shoveling mud from basements, or sheltering neighbors in their homes.

Audio or video documentary is another engaging art medium. As a graduate student, I worked with students to create an audio recording of a semester-long challenge course class. The compilation CD included snippets of the group's dialogue while involved in an intense problem-solving situation and the laughter and celebration as a peer pushed her comfort zone to jump for a trapeze on the challenge course. Another student created a video documenting the group's progress. The process of editing down and choosing which footage to include was a masterpiece of reflection practice.

Experiment with the possibilities of using **performance art** as reflection, and your group may go places you never expected. I've seen some interesting group-closing activities involving short skits, living sculptures, and even a modern dance presentation.

Self-Reflection Techniques

Time spent reflecting alone, away from the group, balances and supports the group process. It gives learners opportunity to reflect on issues that might not have come up during a group discussion or that they might not feel comfortable verbally expressing in a group setting. Opportunities for individual reflection activities away from group time ensure that reflection is happening regardless of what happens during program time. Journaling and other self-reflective methods can become lifelong skills that help develop insight, which is one of the most difficult and most valuable skills to learn in life (Stanchfield, 2001, 2007).

Quiet Solo Reflection

A solo experience that involves time alone in a natural place is historically a traditional part of coming-of-age rites in many cultures throughout the world (Krown, 2009). Allowing participants to take short amounts of time, separate from the group, in a structured and safe way, can be very effective. Participants can be given a journaling or drawing assignment during this time. In a school or therapeutic setting, this might simply involve 10 to 15 minutes of quiet time in the school yard or classroom.

Silent Walk

When I take groups outside for an activity such as exploring our local woods or enjoying art in nature, I sometimes ask participants to separate from each other and walk back to the school silently for some quiet self-reflection.

Mindfulness

Mindfulness is the intentional practice of focusing one's awareness on the present moment, one's emotions, thoughts, and environment. It includes an increased awareness of the body and focusing attention on one's breath. Through mindfulness, participants can learn to calmly acknowledge and accept their thoughts and feelings. It has been widely used as a therapeutic technique for stress reduction and increasing mental clarity, and is now receiving attention as a research-based method for promoting wellness and academic success in schools. In recent years, schools have embraced mindfulness to help improve students' attention, emotion regulation, and learning. There is an emerging international movement in schools and other educational programs promoting mindful awareness that has been proven effective for reducing stress and increasing academic performance in a variety of settings (Gregoire, 2013; Zimmerman, 2013). For more information and resources: www.mindfulschools.org, Mindfulnessinschools.org, MindUPthehawnfoundation.org

Reflective Writing and Journaling

Whether participants are K-12 or adults involved in multi-day programs or one-day experiences, journaling can be an incredibly valuable reflective tool. Reflective writing can involve a variety of mediums including worksheets, drawings, scrapbooking, photography, and audio or video documentary.

Facilitators of school-based and therapeutic programs may have difficulty trying to fit experiential problem-solving activities into a traditional class or group

Michael Thomas, a high school social studies and English teacher pointed out that participants are more likely to stay engaged if you are also quietly writing instead of wandering the room prepping for the next activity.

Jen Bradford allows her students to decide whether or not they want to share their journal entries. They hand in their writing with a "Yes" or "No" written on the folded or stapled page.

One of the highlights of my career involved using journaling as part of the facilitation of a semester-long, high school challenge course class. The value of experiential education—which I intuitively had always known—was reinforced in the students' journal writings. It was an inspiring practice for the students, my co-teacher Donna Richter, and me. There is more about how we used the journaling information as an evaluative, action-research tool in my book, *Tips & Tools: The Art of Experiential Group Facilitation*.

session. Processing is often sacrificed because of this time crunch. Journaling can provide a time for reflection apart from the actual program or class time and gives learners a tangible memoir of their experience and growth.

Learners who journal as part of a long-term program or throughout the school year demonstrate increased insightful thinking and improved writing skills. This has been true in my own ongoing work with people of all ages. An interesting and easy style of journaling that works well for all kinds of programs involves using a sheet of prepared prompts with fun spaces provided for the participant to write/draw their answers. This is far less intimidating than a blank page especially for reluctant or inexperienced writers. A great example for this type of journaling is *The Me I See: Learning Through Writing and Reflection* (see resources).

Some Creative Ways to Implement Journaling

- Make opportunities for participants to create their own journals or scrapbooks. A great resource for bookmaking is *The Bookmaking Experience: An Educator's Guide to Student Made Books* (see resources).
- Include artwork as a part of journaling. Some are more comfortable using artistic representations of their experience.
- Give beginners specific prompts or questions to answer to aid the process.
- Coordinate with the art department of your school or program to collaborate on projects using alternative forms of media for journaling—such as photography and video and audio documentation.

Facilitation Suggestions

- Give students a choice to focus on one of three prompts.
- Journal while they are journaling. Role-model reflective writing when you ask your group to do it.

- The language we use when we introduce an activity is powerful—invite students to participate.
- Emphasize that they don't have to show it to anyone else. They might be asked to share about it, but not necessarily show it to others.

Mementos of Experience: Carrying the Lesson Forward

When learners leave a class or program with an object representing their experience, they physically carry away a piece of the lesson learned. Every time they see this memento it is a tangible reminder of the knowledge gained from the experience. With this in mind, you may want to give participants an opportunity to make or choose something that represents their experience together as a group or is an individual symbol of success. This is an excellent way to infuse learning experiences into day-to-day life (Stanchfield, 2007).

Many of the metaphoric image and object activities described in this chapter can be used as meaningful closing activities and as a gift or memento of experience for learners. I often give quotes, buttons, and Miniature Metaphor charms as a token of experience to participants, especially if they are used in a celebration of strengths activity. Months, even years, after a class or training I will see these items on desks, bulletin boards, or lockers.

Postcard to Your Future Self

Purpose/Focus: reflection, self-reflection, strengths, mindset, goal-setting, 21st century skills, planning, vision setting, memento of experience

This is one of my favorite reflective activities because it encourages learners to reflect on their strengths, celebrate their accomplishments, and clarify their goals, and it gives them a tangible reminder of their learning for the future. On my last day with Amanda Dixon and her 8th grade language arts class, we invited students to pick a postcard that represented their personal strengths. We asked them to think about an important social or life lesson they learned during the past school year that would be important to remind themselves of in September when they reach high school. The students

Postcard examples:
"I learned not to take in everything at once. I should take things one step at a time."

"My commitment is to be myself and not change for others."

"... to work harder, because I don't really work hard for school. In high school it is going to be even more important to try."

"... to be a better listener. When I get overwhelmed, I start talking more than listening, which can annoy others and makes for a stressful day."

"I've learned that being popular and just being yourself are two different things. My hope for next year is to try my hardest and to just be friendly."

The editor-in-chief of a periodical used my postcards for a tribute to her boss, who was retiring from an organization that he founded 40 years ago. She asked board and staff members to pick a card that brought to mind one thing their boss had taught them or that exemplified him. She photographed the postcards for a PowerPoint® presentation. As each slide came up, the person who picked the image shared why they had picked it. It worked wonderfully. She planned to follow up by asking them to write a note on the back of their cards to send to him over the next few weeks.

then wrote a note addressed to themselves that included this lesson and a commitment for the future. After students finished writing, we met with them individually to talk about the lessons learned and personal commitment as an exit interview for the year. We sent the cards to the students in early September at the start of their 9th grade year. The two questions we asked students to answer:

- What is an important lesson you learned in middle school that you don't want to forget, and that you think will help you in the future?
- What is one commitment you are willing to make to your 9th grade self?

This has now become a yearly ritual. We start a month prior, using the Trade and Share activity during social studies class to celebrate the strengths of their peers (see page 170). Then the last week of school we pull the cards out again for this activity and an exit interview with each student.

I use this as a closing activity in my workshops with adults. At the end of a workshop or course, participants write down at least one thing they learned from the workshop activities or through a conversation with a peer that they don't want to forget. I then ask them to write a commitment to themselves around changing, improving, or trying something new in their practice as an educator. Many workshop and training attendees later report to me that the card arrived on "just the right day" to remind them of a personal strength, lesson, or goal.

Principles of Effective Reflection

- Reflection is a practice that takes time to develop in individuals and groups. Be conscious of sequencing and scaffolding reflective activities. Start with simple choice and partner-dialogue activities before asking participants to share in a large group setting.
- Be mindful of your attitude toward facilitating reflection. How do you approach this aspect of teaching and facilitation? Learners respond to your cues. If you approach reflection as an exciting and enjoyable activity rather than chore, participants will likely perceive it positively too.
- Be open to experimentation, and learn from your participants. Learner-centered methods can add fun and interest to reflection and great value and meaning to group experiences.
- Promote inquiry and exploration over interpretation. Help participants take ownership and control of interpreting and applying the experience to their lives and to future learning, rather than interpreting the experience for them.
- Help learners perceive the similarities or "patterns" between the learning experience and other aspects of learning and life.
- Reflection begins when the group comes through the door and can be blended throughout a program. There is no one right way or one right time to facilitate reflection.
- Create a physical environment, room and group setup where people can see and hear each other. Be cognizant of possible distractions such as temperature, noise, time of day, seating arrangements, and so on.
- Allow group members to pass during discussions. Ask for volunteers rather than call on individuals. This empowers learners to take control of their learning and practice sharing at their own pace. The power to choose how and when one shares promotes trust in the facilitator and the process.
- Be prepared for the group to take discussions somewhere different than you had in mind. You might learn something new! Be aware that you might have to artfully help them navigate back to the present moment in order to meet the needs of the group.
- The facilitator doesn't have to hear everything said in group reflection activities for it to be effective. Regularly blend in activities that don't involve you directly in the discussion by dividing the group into smaller reflection groups or active pair share activities. Trust that you and the rest of the group will receive the information you need through group process.
- Silences are okay—even necessary. Take time for people to think and respond.
- Participants can be experiencing valuable reflection even if they don't share it with the group.
- The more ways a group or individual reflects or processed the more ways they will be able to recall and apply the information.
- There is a lot of power and emotion in the moment. Take advantage of teachable moments as they arise.

- Be intentional about planning reflection and plan your reflective activities as you do the lessons and activities—and then be willing to let go when other unplanned teachable moments arise.
- Promote the importance of active listening and being fully present with group members.
- Direct questions back the group, letting group members help each other.
- Use a variety of reflective methods—group dialogue, active reflection methods, artwork and tangible objects, images, or metaphors.
- Closure is important. Be thoughtful about how you end a program or a day's experience. Design a closing activity that helps the group tie their learning together and encourages future learning. Bookend.

Reflection can be a dynamic, engaging, even enjoyable part of teaching and facilitation. Now you have a full toolbox of innovative techniques to turn those teachable moments into profound insights and lasting lessons. These methods will help you meet the diverse needs of the participants in your groups and take advantage of multiple pathways to increase relevancy, depth of understanding, and lasting lessons.

Chapter ten picks up on this theme of reflection and focuses on another key ingredient to successful teaching—practitioner reflection. Taking time to reflect on one's practice to improve outcomes, and ultimately, satisfaction with work is foundational to being an inspired educator.

chapter 10

The Reflective Educator

Meaningful Self-Reflection and Record Keeping to Improve Your Practice

We must, in other words, become adept at learning. We must become able not only to transform our institutions, in response to changing situations and requirements; we must invent and develop institutions which are 'learning systems,' that is to say, systems capable of bringing about their own continuing transformation. —Donald A. Schön, *Beyond the Stable State* (1973)

Educators spend so much time focusing our energy outwardly. We spend our days planning, preparing, meeting, teaching, and implementing. Many of us work in fast paced environments and spend a lot of time rushing from one place to the next, or working with one group after another in our classroom. Often, when one program or class ends, we move onto the next without spending time to reflect on our successes or the lessons we could be taking away to improve our practice. The philosophy of experiential education recognizes reflection as the key to moving learning forward and creating lasting and meaningful lessons. We seek out new tools and ideas to facilitate reflective practice with our students or groups, but how often do we practice what we preach and take time for self-reflection?

Many of us find that daily reflection and record keeping can be a great challenge. Often we engage only in record keeping that is required for administrative needs rather than meaningful reflection that will improve our practice. Taking the time to reflect on our work helps us find meaning, and develop insight into what strategies or approaches are most effective. It helps us use what we learn each day from our clients or students to improve in the future.

As we explored in chapter four (pages 48-49), reflective practice is an important part of documenting and communicating the value of your program or course offerings. Ongoing gathering of data such as quotes, student work, and client testimonials can help you "tell the story" of your program or class and monitor progress and successes along the way. Maintaining an ongoing record of lesson plans and activity sequences can be very useful in planning future programs or lessons. Some of our best tools and strategies are developed "on the fly," when a lesson plan goes awry or we adapt to the needs at hand and create something new or better. If we don't have a systematic way to capture these teachable moments they can be lost.

There are some practical ways to start making record-keeping a habit. Some of the possibilities include an electronic or handwritten journal, a daily lesson or training report, a scrapbook/sketchbook, a space on your lesson planning sheet or calendar for notes, an activity log, or regular check-ins and report-outs with a co-teacher or colleague. Some educators keep electronic notes regularly throughout their day on their phone or tablet. Experiment and make it meaningful and useful for you.

The ongoing logbook of lesson plans and outcomes I keep about my work includes notes on successes, variations of activities and program outcomes. I alternate between handwritten and electronic versions of my notes and include pictures, quotes from group members, group generated lists, drawings, and other artifacts. This logbook has helped me understand what activities work best and when. I use it when the educators and trainers I work with want follow-up resources and notes. It helps facilitate our planning of what's next, improving upon lessons, initiating creative inspiration for new activities, and identifying opportunities for collaboration.

These records have also allowed us to see the progress of our students over time and share the value of these experiences to administrators, parents, other educators, and community members. Regardless of how you keep records, the most important thing is to make reflection a regular habit and to record not only what happened but also what it meant and how the information will improve your work.

Suggested Questions for Daily Reflective Practice

- What was the goal of the lesson or experience? Was that goal met? How?
- Did you learn something new about the activity? Were there some new adaptations or adjustments to the lesson that should be used again?
- How did the group progress through the activities/material? Was there enough time?
- What were some participant reactions to or reflections on the experience?
- What did you learn from the participants? What did you learn about yourself?
- Was there an important moment or outcome that should be shared with a program administrator or other interested party?
- Are there photos, testimonials, quotes, journal statements or other documentation of the experience and outcomes that should be saved or shared?
- What would you do differently in the future?
- What will you do to follow up the next time the group or class meets?

It is also important to take time to reflect on the "bigger picture" of your practice as an educator including your personal and professional goals, accomplishments and next steps. This kind of self-reflection not only helps you improve your work, it can also make it more meaningful and rewarding. Reflection on why you do what you do and the successes and opportunities that have arisen along the way increase your focus and sense of purpose.

Suggested Questions About Your Ongoing Practice as an Educator

- Why do you do what you do? Why do you keep doing it? What brings joy in your work?
- Identify a professional or personal success or achievement you have made over the past year, week, month. What made this a success? How did you do it?
- What is a "success story" with a student that you would share with a friend or colleague?
- What does success mean to you? To your students or clients?
- Reflect on a strength you bring to your work. How do you maintain or capitalize on that strength?
- When are you most effective in your practice of facilitation or teaching? What are the components of a great day of teaching or training?
- What is a personal or professional challenge for you right now?
- What have you done to face these challenges? What will you continue to do?
- What will you do differently in the next month? Who can you call upon for support if you need it? What professional development opportunities do you want or need to seek out in the near future?
- What is a lasting lesson from your experiences as an educator this month? How will you remind yourself of this?
- How can you share your learning with someone else? Who can you collaborate with to improve your work and your enjoyment of work?

Consciously carving out time in your schedule to reflect on your students' growth, successes, and needs will directly affect the methods and strategies you use. It will connect you to your own strengths, needs, and goals and help you make better decisions.

The key is to make regular time for this practice—even if it is just a short time each day or a longer period each week. For me, the early morning hours, when the house is quiet, before the phone starts to ring, or before I have to think about my first appointment, is the ideal time. The clarity and focus I can achieve during the solitude of this time of day is well worth getting out of bed earlier.

Self-reflection and communication with other educators about your work helps you develop a clearer purpose, meaning and fulfillment in your practice as an educator. You will find that when you regularly reflect on your work and its meaning, methods and outcomes, you will continually improve your practice by looking more deeply into the WHY behind what you do.

Experiential Lesson Planning Questionnaire

"What makes a lesson plan experiential? How do I design an experiential lesson plan?" In my graduate courses, I've found a number of educators asking for guidelines on how to create an experiential lesson plan. When asked this question, I found myself naturally referring back to the principles of experiential education and brain-based learning. During one of these conversations in a recent graduate course, the teachers I worked with helped me put a checklist together based on the reflective questions I asked them about their lessons. Thank you to Amanda Klein from Lincoln-Sudbury Regional High School for putting the questions into a checklist form.

Questions Educators Should Consider When Designing Lessons

- Does the lesson include a way to facilitate a clear understanding of goals and outcomes of the lesson/unit?
- How will you help students understand the goals and outcomes?
- What is your hook to engage learners in this lesson?
- How are you creating interest in the subject?
- How is the subject meaningful and relevant to student's day-to-day or future life? How are you going to help them see this relevancy?
- What opportunities are you giving learners to take control and ownership over their learning?
- How are you using multiple pathways to learning? Does the lesson involve multiple senses? Are you including activities that are physical and social? Is there an emotional connection to the material? Are you using multiple approaches such as movement, art, metaphor, writing and dialogue to deliver the lesson?
- What techniques are you using to take advantage of transitions, i.e., beginning and ending of class (primacy/recency) and after direct instruction.
- How are you cultivating a positive community and building positive relationships with your students?
- What 21st century, social-emotional skills are you promoting? How?
- Are there incremental, achievable challenges in your lesson/unit design?
- How have you sequenced or scaffolded the lesson?
- What cross-curricular connections can you make? Is this an opportunity to collaborate with other educators, community members, and/or parents?
- What role can students take in leading activities as part of this lesson?
- What types of evaluation and formative assessment are you using throughout this lesson?

Experiential Lesson Planning Checklist 

Technique	Fully	Partly	No	Activity/Strategy
Clear Goals and Outcomes Will learners understand the goals and purposes?				
Hook Will learners engage/buy-in at the beginning?				
Meaning and Relevancy Will learners see the relevancy to their everyday life?				
Student-Centered Can learners take ownership over their learning?				
Multiple Pathways Are multiple strategies (i.e., physical, social, emotional, movement, art) being used to connect to the learning?				
Engagement and Reflection Will students connect to the learning at the beginning and ending of class, during transitions, and after direct instruction?				
Community-Building Will learners build positive relationships/ community?				
21st Century/Social-Emotional Skills Will learners build skills with the learning?				
Formative Assessment Are learners using reflection, feedback and review to connect with the learning?				
Sequencing/Scaffolding Are there incremental and achievable challenges in the learning?				
Cross-Curricular Are there cross-curricular connection?				

 Adapted from the collaborative creation of the Inspired Educator, Inspired Learner 2013 summer class. Facilitator: Jen Stanchfield. Participants: Greg Gammons, Lori Hodin, Amanda Klein, Beth Mangano, Marci Stoda, Sam Parker, Amy Thivierge, and Susan Totten.

The Practice of Teaching

My goal in putting together this book was to share the techniques I have found successful in my own practice so that others might benefit from my experiences and find some valuable tools to add to their teaching and group facilitation toolbox. It is my hope that readers have gained new perspectives about the intertwined pedagogical theory and practice of experiential education, differentiation, 21st century learning, social-emotional skills development, and the exciting new research on the brain and learning that gives scientific support for these practices.

My intent is to inspire you to reflect on your practice as an educator and what you love about teaching. Effective teaching is a blend of art, science, and most importantly, practice and commitment. Educators can develop these skills through ongoing practice, experimentation, training, collaboration, evaluation, delving into theory, learning from other educators, and learning from their students. Successful educators take into consideration the unique strengths and personalities that exist in each group with the intention of meeting the differing needs of learners. This is a dynamic, ever-changing growth process for educator and learner. My hope is this book is a useful tool in supporting this process.

Acknowledgments

More people than I could ever name contributed to this book. It is a reflection of 25 years spent as an experiential educator with amazing groups of students, group facilitators, and teachers in a variety of educational settings. My sincere thanks to the wonderful programs and individuals I have been fortunate enough to work with over the years. You inspired this book, and the lessons I learned from our experiences are the foundation of this book.

Thank you to managing editor Mony Cunningham of Wood N Barnes Publishing Company for believing in this book from the beginning and never wavering in your support throughout the years it took to pull it together. You provided the wisdom and patience necessary to bring this book to fruition and make it reader-friendly. Working with you has been a highlight of my career.

I am grateful to David Lockett, principal of Jefferson Elementary School in Stevens Point Wisconsin, who provided the most amazing opportunity to develop my craft and passion for bringing experiential education practices into public schools. You inspired me with your vision for integrating experiential principles into day-to-day practice in education and your deep interest in neuroscience and its application to teaching.

Lynn Reining, Nancy Wyngaard, Donna Richter, Betsy Beuchner of Middleton-Cross Plains area school district provided me with incredible opportunities to cultivate my teaching skills and implement new activities, ideas, and methods with their teachers and students. Thank you for inviting me to collaborate in your classrooms and teacher professional development programs. To Polly Chandler, Joe Boggio, Patrice Strifert and the teachers and students at Hinsdale Middle and High School, thank you for the opportunity to learn and grow with you during our time together with the Antioch Co-Seed project.

A special thank you to all of the teachers and staff at Twin Valley Middle School for allowing me to be part of your teaching team. Thank you to Bill Anton for first writing me into the school's action plan, and to Marie Paige for hiring me to provide professional development programming. Thank you Jessica Hammond, Amanda Dixon, Anne Sulzmann, Scott Salway, Yolanda D'Allesio, Angel Colford, Keli Gould, Suzanne Mears, Tracey Gamache, Karen Horton, Samantha Ambrose, Francesca Palumbo, Sandy Crowningshield, and Keith Lyman for continued collaboration and the opportunity to explore the application of experiential teaching methods in TVMS classrooms. I learn so much every day from you and your students.

Thank you to Mt. Anthony Union High School teachers and staff, especially Molly McWithey-Maish, Amy Thivierge, Mike DiMaio, Jessica Dubie, Mark Upright, Sue Maguire, Paul Reif, and Jean Zinn. I appreciate your co-creative work on developing the advisory programming and sharing your feedback on experiential strategies used in your classrooms. To Johanna Liskowsky-Doak, Amanda Dixon, and the Leland & Gray Middle School team for inviting me to provide ongoing programming with your students and their parents. It is a pleasure to work with students and families in my own community. I appreciate your willingness to experiment and learn and grow together.

To Sam Wang and Judy Willis, thank you for inspiring me with your knowledge and accessible style of presenting the latest research on the brain and learning. Your work has been integral in bridging the gap between educators and neuroscientists.

Thank you to Sandy Negley, Amy Long, Mauria Tanner, Jeri Claspill, Greg Smith, Steve Bell, Patricia Bishop, Kelly Mattingly and the other talented members of the Expressive Therapies Department at the University of Utah Neuropsychiatric Institute for teaching me a multitude of tools and tricks for facilitating groups and weaving important life lessons into experiential group work in the classroom and therapeutic group room.

To Dr. Jasper Hunt, thank you for reintroducing me to the work of John Dewey and inspiring deeper study into pedagogical philosophy. You reinforced the value of understanding the "why" behind what we do in experiential education and helped me connect the dots between theory and practice. Thank you Dr. Steven Simpson of the University of Wisconsin LaCrosse for helping me define my own philosophy of education, introducing me to valuable insights from eastern philosophy, and encouraging me to present and share my work with others in the field.

Thank you to Ms. Janet Stoddard, my 5th grade teacher, for your example of caring and committed teaching and your ability to know, understand, engage, and motivate the individual students in your classroom and for inspiring me to become an educator.

Thank you to Marci Charles of Team BRIDGES Memphis, Valerie Bodet, Matt May and Traci Salisbury of LOOP New Orleans for sharing photos from your programs. Carrie Lane and Sandy Crowningshield, your help with photo identification and permissions is greatly appreciated. To Karl Rohnke, Kara Bixby, Brenda Reed, Carol Carlin, Emily Case, and Kathy Haras, thank you for making the time to edit and support this project.

Finally, to my husband Paul, I appreciate your encouragement, patience, and willingness to cheerlead, edit, and critique throughout the years spent growing this book.

Activity List/Alphabetized

Activity	Page #	Purpose
Active Coordinate Plane	127	math, academic review/application, formative assessment, collaboration, problem-solving
Alaskan Baseball	184	fair play, following directions, taking turns, high activity game, community building
Anyone Who (Adapted from Have You Ever?)	98 205	academic review, formative assessment, reflection, group discussion, community building, collaboration, context setting, commonalities/different perspectives, turn taking
Appreciation Notes	168	celebrations/appreciations, descriptive language, reflection, strengths, trust, mindset, group norms, citizenship
Attention Getters	61-3	organizing groups, transitions, focus, self-regulation, following directions
Back Art	157	communication, salient points, focus, fair play, appropriate touch, community building
Back-to-Back Drawing	158	communication, patience, focus, giving/receiving directions, leadership/followership
Bag of Nouns	114	academic review, community building, descriptive language, active listening, formative assessment, creativity/imagination, focus, turn taking, self-regulation, fair play
Charades Race	112 206	academic review, formative assessment, vocabulary, reflection, communication, community building, creative thinking, turn taking, self-regulation, focus, fair play, energizer, innovation
Circle 'Em Up	63	organizing groups, attention getter, collaboration, social-emotional, following directions
Classroom Four Square	102	academic review, formative assessment, self regulation, turn taking, fair play, quick energizer
Collaborative List Making Review	128	academic review, collaboration, formative assessment, active listening, vocabulary, collaboration, turn taking, fair play, self-regulation, focus
Colleen's Character/Metaphor Lesson	124	language arts, reading comprehension, metaphor, discussion/writing prompt, reflection, descriptive language, formative assessment, content application, creativity/imagination, communication, emotional connection to learning, empathy
Coming and Going of the Rain	191	collaboration, active listening, focus, taking turns, communication, self-regulation, quick energizer
Commonalities Mingle	97 206	context setting, celebrations/appreciations, community building, conversation starter, reflection, communication, commonalities/different perspectives, formative assessment, energizer, organizing/dividing groups, choice/control/ownership
Communication Breakdown	155	communication, frustration, giving/receiving directions, active listening, community building, trust, leadership/followership
Community Circles	151	compromise, conflict resolution, community building, creativity, active listening, critical thinking, following directions, innovation, problem-solving, communication, leadership, responsibility, appropriate touch, big picture perspective, group norms
Community Puzzle Pieces	173	community building, strengths, mindset, reflection, celebrations/appreciations, group norms, creativity/imagination
Computer Keyboard Keys	77 203	hook, context setting, "first five" or "do now," reflection, community building, metaphor, descriptive language, discussion/writing prompt, creativity/imagination
Concentric Circles	86 205	academic review, formative assessment, group discussion, active listening, names, community building, reflection, commonalities/different perspectives, energizer, transition

Activity List (continued)

Activity	Page #	Purpose
"Conversation Starter" Pin-Back Buttons	76 201	hook, context setting, reflection, emotional connection to learning, community building, discussion prompt, choice/control/ownership, self-expression
Create a Colony	123	formative assessment, content application, community building, group norms, communication, planning, decision-making, vocabulary, creativity/imagination, citizenship
Daily Riddle or Brainteaser	80	hook, "first five" or "do now," transition, conversation starter, creativity, critical thinking, problem-solving
Dividing Groups	64-5	organizing groups, playful learning, following directions, active engagement
Dominos and Other Simple Prop "Match Ups"	95 205	dividing groups, discussion, reflection, community building, academic review, formative assessment, math, problem-solving, commonalities/different perspectives, transition, names, choice/control/ownership
Dominos Fraction Line Up	118	academic review, formative assessment, math, collaboration, communication, problem-solving, leadership/followership, active listening, big picture perspective
Draw Your School/Workplace	174	group norms, goal setting, community building, reflection, metaphor/imagery, citizenship, collaboration, vision setting, responsibilities, creativity/imagination, communication
Elbow Tag	186	high activity/field/gymnasium game, community building, appropriate touch, boundaries, turn taking, responsibility, self-regulation
Fill the Crate	158	communication, collaboration, problem-solving, creativity/imagination, using resources, planning, persistence, community building, critical/higher order thinking, leadership/followership, patience/managing frustration
"Follow Me"	154	impact of actions, leadership/followership, peer pressure, collaboration, group norms, discussion prompt, quick energizer, transition, responsibility
Food Web Tag	126	science, academic review/application, formative assessment, community building, fair play, self-regulation, appropriate touch, big picture perspective, high activity game, vocabulary
Gallery Walk	104	collaboration, reflection, review, goal setting, formative assessment, conversation starter, commonalities/different perspectives, vision setting, transition, community building
Gotcha	192	community building, appropriate touch, quick energizer, focus, following directions, trust, transitional, collaboration
Graffiti Wall	83 213	hook, emotional connection to content, "first five" or "do now," transition, conversation starter, community building, empathy, discussion/writing prompt, reflection, collaboration, creativity/imagination, group norms
Group Juggle Metaphor	177	collaboration, planning, communication, problem-solving, time management, coping, stress management, identifying resources, reflection, empathy, group norms, discussion prompt
Group Norms Sculpture/Symbol	175-6	group norms, reflection, collaboration, citizenship, responsibilities, metaphor, creativity/imagination, communication, goal setting
Group Sculpture for Academic Review	82	hook, "first five" or "do now," reflection, academic/content review, formative assessment, metaphor, collaboration, creativity/imagination, problem-solving

Activity List (continued)

Activity	Page #	Purpose
Handshake Mingle (Names Emphasis)	89 143 204	group discussion, community building, academic review, reflection, communication, choice/control/ownership, names, commonalities/different perspectives, appropriate touch, transition, quick energizer, conversation starter
Have You Ever?	140	active listening, community building, commonalities/different perspectives, self-regulation, turn taking, reflection, academic review, formative assessment, celebrations/appreciations
Headlines	183	goal setting, values clarification, reflection, mindset, planning, creativity, communication
Helium Hoop	162	communication, collaboration, problem-solving, focus, patience/managing frustration, synergy, persistence, leadership/followership
How Long is a Minute?	182	time management, writing/discussion prompt, perceptions vs. reality, self-regulation, reflection, quick energizer
Images/Postcards	71	hook, "first five" or "do now," conversation starter, context setting, discussion/writing prompt, reflection, community building, figurative language, metaphor, emotional connection to learning, creativity/imagination, choice/control/ownership, group norms
Invisible Clay	185	creativity/imagination, communication, collaboration, focus, innovation, turn taking, transitions, imagery, community building
Journaling or Reflective Writing Prompts	79 204	context setting, "first five" or "do now," reflection, review, transition, creativity/imagination, communication, self-expression, formative assessment
King Frog/Queen Elephant/ Animal Kingdom Game	190	community building, focus, communication, transitions, fair play, quick energizer
Layers of the Atmosphere (and Other Sequence Challenges)	117	academic review, formative assessment, sequencing, reflection, differing perspectives, active listening, focus, patience/managing frustration, big picture perspective, problem-solving, leadership/followership, collaboration, critical/higher order thinking
Musical Connection	82	hook, attention getter, transition, context setting, introducing a topic, emotional connection to learning
Mystery Leader	188	transitions, self-regulation, following directions, leadership/followership, focus, taking turns, collaboration
Name Guess	147	names, community building, active listening
Name Meanings	143	community building, names, respectful behavior, group norms, celebrating differences, trust, communication
Name Roulette	145	names, fair play
Newspaper Bridges	161	collaboration, problem-solving, communication, using resources, innovation, planning
Pass It To	147	name reinforcer, community building, self-regulation, taking turns
Pass the Knot	103	academic review, reflection, group discussion, formative assessment, community building
Peek-a-Who (Names or Complements)	145, 166, 207	community building, name practice, empathy, taking turns, communication, honoring individuals, celebrating differences, commonalities/different perspectives, fair play

Activity List (continued)

Activity	Page #	Purpose
Play Dough Pictionary	110 207	academic review, formative assessment, vocabulary, reflection, communication, collaboration, creativity/imagination, self-regulation, fair play, focus, turn taking, energizer, critical/higher order thinking
Postcard Strength	169	celebrations/appreciations, strengths, mindset, descriptive language, metaphor, reflection, community building, citizenship, self-expression, group norms, writing prompt
Postcard to Your Future Self	217	reflection, self-reflection, strengths, mindset, goal-setting, planning, memento of experience
Question or Conversation Starter of the Day	81	hook, "first five" or "do now," conversation starter, context setting, community building, reflection
Quotes	75 202	hook, "first five" or "do now," context setting, conversation starter, reflection, writing prompt, emotional connection to learning, formative assessment, community building, transition
Reflection Activities	201-218	emotional connection to learning, self-regulation, coping, stress management, focus, strengths, mindset, goal setting, communication, strengths, creativity/imagination, metaphor, self-reflection, planning, memento of experience
Review and Reflection Dice	104 208	academic review, reflection, conversation starter, formative assessment, group discussion, vocabulary, alternative to question answer session
River Crossing (Math/Insightful Problem-Solving)	121	collaboration, math skills, logic, problem solving, critical/higher order thinking, communication, patience/managing frustration, big picture perspective, persistence
Rock-Paper-Scissors Tag	149	collaboration, communication, conflict resolution, consensus, compromise, decision-making, listening, leadership/followership, frustration, fair play, community building
Run and Scream	192	high activity/gymnasium/field game, quick energizer, transition
Saving Springfield	163	problem-solving, communication, managing frustration, leadership/followership, decision-making, collaboration, creativity/imagination, using resources, critical thinking, community building, giving/receiving directions
Self-Reflection Activities	214	reflection, self-regulation, critical thinking, introspection
Smaug's Jewels	189	self-regulation, turn taking, collaboration, fair play, appropriate touch
Team Tally	135	transition, conversation starter, community building, communication, commonalities/different perspectives, reflection, fair play, collaboration, context setting, academic review, formative assessment, discussion technique, emotional connection to learning, empathy
Telegraph	152	communication, self-regulation, focus, trust, gossip/rumors, appropriate touch, fair play, collaboration
Telegraph and Probability	120	math skills, probability, academic review, communication, trust, collaboration, self-regulation, appropriate touch, fair play, gossip/rumors, focus, community building
Time Management "Full Plate"	180	time management, goal-setting, prioritizing goals, coping, self-regulation, stress management
Tin Can Pass	148	communication, collaboration, problem-solving, patience, creativity/imagination, active listening, goal setting, giving/receiving directions, trust, planning, leadership/followership

Activity List (continued)

Activity	Page #	Purpose
Toolbox, Object, or Miniature Metaphor Charms	73 203	hook, context setting, discussion/writing prompt, "first five" or "do now," conversation starter, reflection, community building, figurative language, metaphor, emotional connection to learning, creativity/imagination, choice/control/ownership, group norms
Trade and Share (Names Focus, Strengths, Review)	91, 146, 170, 205	social-emotional learning, communication, active listening, community building, reinforce names, formative assessment, reflection, multiple pathways to learning, academic review, reflection, mindset, celebrations/appreciations, descriptive language, metaphor
Triangle Tag	187	cooperation, collaboration, fair play, self-regulation, turn taking, appropriate touch
Turning Over a New Leaf (aka Magic Carpet)	160	communication, collaboration, creative thinking, problem-solving, trust, spatial relations, innovation, appropriate touch, planning, persistence, patience, leadership/followership
Turnstiles	159	communication, problem-solving, collaboration, taking risks, planning, using resources, community building, followership/leadership, trust, persistence, gymnasium/field game
US List	50	group norms, discussion prompt, community building, citizenship, self-regulation, responsibility, collaboration, communication, choice/control/ownership, empathy, gossip/rumors, goal-setting, trust, emotional connection to learning, mindset
Vintage Keys "Key to Our Group's Success"	171	group norms, reflection, discussion/writing prompt, community building, citizenship, responsibility, strengths, mindset
Which One?	65	dividing groups, conversation starter, communication, compromise, collaboration, decision-making, conflict resolution
Who Has It?	189	community building, self-regulation, turn taking, fair play, collaboration
Zoom	115 156	community building, problem-solving, communication, leadership/followership, managing frustration/patience, collaboration, different perspectives, higher order/critical thinking, active listening, big picture perspective, descriptive language, problem-solving, focus, sequencing, persistence

All of the activities in this book exemplify differentiated instruction, strength based teaching methods and ways to promote 21st century skills and social-emotional learning. This list is a sampling of activities in upcoming chapters will help you reach these goals.

Purpose List

Purpose	Activities		
Active Listening	Concentric Circles Trade and Share Bag of Nouns Zoom Layers of the Atmosphere Dominos Fraction Line Up	Collaborative List Making Have You Ever? Name Guess Tin Can Pass Rock Paper Scissors Tag Community Circles	Communication Breakdown Celebration/Appreciation Personal Strengths Activities Reflection Activities
Alternative Discussion Techniques	Concentric Circles Handshake Mingle Trade and Share Dominos/Match Ups	Commonalities Mingle Anyone Who Pass the Knot Review and Reflection Dice	Gallery Walk Team Tally
Appropriate Touch/ Boundaries	Handshake Mingle Telegraph and Probability Food Web Tag	Handshake Mingle Community Circles Telegraph	Turning Over a New Leaf Large Group/After School Transition/Filler Activities
Attention Getters	Circle 'Em Up	Musical Connection	
Big Picture Perspective/ Sequencing	Zoom Layers of the Atmosphere	Dominos Fraction Line Up River Crossing Challenge	Food Web Tag Community Circles
Commonalities/ Different Perspectives	Concentric Circles Handshake Mingle Trade and Share Dominos/Match Ups Commonalities Mingle	Anyone Who Gallery Walk Zoom Layers of the Atmosphere Team Tally	Have You Ever? Handshake Mingle Peek-a-Who Names Reflection Activities
Communication	US List Which One? Journaling/Writing Prompts Concentric Circles Handshake Mingle Trade and Share Commonalities Mingle Play Dough Pictionary Charades Race Bag of Nouns Layers of the Atmosphere Dominos Fraction Line Up Telegraph and Probability	River Crossing Challenge Create a Colony Colleen's Lesson Team Tally Have You Ever? Handshake Mingle Name Meanings Peek-a-Who Names Tin Can Pass Rock Paper Scissors Tag Community Circles Telegraph Communication Breakdown	Zoom Back Art Back-to-Back Drawing Fill the Crate Turnstile Turning Over a New Leaf Newspaper Bridges Helium Hoop Saving Springfield Personal Strengths Activities Time Management Activities Reflection Activities
Community Building/ Collaboration	US List Images/Postcards Toolbox/Metaphors Quotes Conversation Buttons Computer Keyboard Keys Question of the Day Group Sculpture Graffiti Wall Concentric Circles Handshake Mingle Trade and Share Dominos/Match Ups	Commonalities Mingle Anyone Who Pass the Knot Gallery Walk Play Dough Pictionary Charades Race Bag of Nouns Zoom Layers of the Atmosphere Dominos Fraction Line Up Telegraph and Probability Create a Colony Food Web Tag	Team Tally Have You Ever? Handshake Mingle Name Meanings Peek-a-Who Names Pass It To Name Guess Tin Can Pass Rock Paper Scissors Tag Community Circles Communication Breakdown Back Art Fill the Crate

Purpose List (continued)

Purpose	Activities		
Community Building/ Collaboration (con't)	Turnstile Saving Springfield Personal Strengths	Time Management Large Group/After School	Transition/Filler Reflection Activities
Conflict Resolution/ Compromise	Which One? Rock Paper Scissors Tag	Community Circles	Reflection Activities
Context Setting/ Introducing a Topic	Images/Postcards Quotes Conversation Buttons	Computer Keyboard Keys Journaling/Writing Prompts Question of the Day	Musical Connection Commonalities Mingle Anyone Who
Conversation Starter/ Writing Prompt	US List, Which One? Images/Postcards Toolbox/Metaphors Quotes Conversation Buttons Computer Keyboard Keys Daily Riddle/Brainteaser Question of the Day Graffiti Wall	Concentric Circles Handshake Mingle Trade and Share Dominos/Match Ups Commonalities Mingle Anyone Who Pass the Knot Review/Reflection Dice Gallery Walk	Zoom Colleen's Lesson Team Tally Have You Ever? Personal Strengths Activities Time Management Activities Reflection Activities
Creativity/Imagination	Images/Postcards Toolbox/Metaphors Computer Keyboard Keys Journaling/Writing Prompts Daily Riddle/Brainteaser Group Sculpture Graffiti Wall	Play Dough Pictionary Charades Race Bag of Nouns Zoom Create a Colony Colleen's Lesson Tin Can Pass	Community Circles Fill the Crate Turning Over a New Leaf Newspaper Bridges Saving Springfield Personal Strengths Activities Reflection Activities
Critical/Higher Order Thinking	Daily Riddle/Brainteaser Play Dough Pictionary Zoom	Layers of the Atmosphere River Crossing Challenge Community Circles	Fill the Crate Saving Springfield
Descriptive Language	Images/Postcards Toolbox/Metaphors Computer Keyboard Keys Review/Reflection Dice Play Dough Pictionary	Charades Race Bag of Nouns Zoom Create a Colony Colleen's Lesson	Celebration/Appreciation Personal Strengths Reflection Activities
Empathy/Citizenship	US List, Graffiti Wall Colleen's Character and Metaphor Lesson	Create a Colony Celebration/Appreciation Personal Strengths Activities	Team Tally Peek-a-Who Names Reflection Activities
Energizer/Transition	Quotes Journaling/Writing Prompts Daily Riddle/Brainteaser Music Connection Graffiti Wall Concentric Circles	Handshake Mingle Commonalities Mingle Classroom Four Square Play Dough Pictionary Charades Race Dominos/Match Ups	Gallery Walk Team Tally "Follow Me" Transition/Filler Activities
Fair Play/Turn Taking	Anyone Who Classroom Four Square Play Dough Pictionary Charades Race Bag of Nouns Telegraph and Probability	Food Web Tag Collaborative List Making Team Tally Have You Ever? Name Roulette Peek-a-Who Names	Pass It To Rock Paper Scissors Tag Telegraph Back Art Large Group/After School Transition/Filler Activities

Purpose List (continued)

Purpose	Activity		
Formative Assessment/ Differentiation	Quotes Journaling/Writing Prompts Group Sculpture Concentric Circles Trade and Share Dominos/Match Ups Commonalities Mingle Anyone Who	Classroom Four Square Pass the Knot Review/Reflection Dice Gallery Walk Play Dough Pictionary Charades Race Bag of Nouns Layers of the Atmosphere	Dominos Fraction Line Up Create a Colony Colleen's Lesson Food Web Tag Active Coordinate Plane Collaborative List Making Team Tally Reflection Activities
Giving/Following/ Receiving Directions	Circle 'Em Up Tin Can Pass Community Circles	Communication Breakdown Back-to-Back Drawing Saving Springfield	Large Group/After School Transition/Filler Activities
Goal-Setting/Planning/ Using Resources	US List Gallery Walk Create a Colony Tin Can Pass	Fill the Crate Turnstile Turning Over a New Leaf Newspaper Bridges	Saving Springfield Group Norms Activities Time Managements Reflection Activities
Group Norms	US List Images/Postcards Toolbox/Metaphors Graffiti Wall	Create a Colony Name Meanings Community Circles "Follow Me"	Celebration/Appreciation Personal Strengths Group Norms Activities Reflection Activities
High Physical Game	Food Web Tag, Turnstile	Large Group/After School	
Hook, "First Five" or "Do Now"	Images/Postcards Toolbox/Metaphor Quotes Conversation Buttons	Computer Keyboard Keys Journaling/Writing Prompts Daily Riddle/Brainteaser Question of the Day	Group Sculpture Music Connection Graffiti Wall
Leadership/ Followership	Zoom Layers of the Atmosphere Dominos Fraction Line Up Tin Can Pass Rock Paper Scissors Tag Community Circles	"Follow Me" Communication Breakdown Back-to-Back Drawing Fill the Crate Turnstile Turning Over a New Leaf	Helium Hoop Saving Springfield Transition/Filler Activities Reflection Activities
Math/Science	Dominos/Match Ups Layers of the Atmosphere Dominos Fraction Line Up	Telegraph and Probability River Crossing Challenge	Active Coordinate Plane Food Web Tag
Metaphor/ Figurative Language	Images/Postcards Toolbox/Metaphors Computer Keyboard Keys	Group Sculpture Colleen's Lesson Personal Strengths Activities	Group Norms Activities Reflection Activities
Names, Learn/ Reinforce	Concentric Circles Handshake Mingle Trade and Share	Dominos/Match Ups Name Meanings Name Roulette	Peek-a-Who Names Pass It To Name Guess
Organizing/ Dividing Groups	Circle 'Em Up Which One?	Dominos/Match Ups	Commonalities Mingle
Patience/Managing Frustration	Zoom Layers of the Atmosphere River Crossing Challenge Tin Can Pass	Rock Paper Scissors Tag Communication Breakdown Back-to-Back Drawing Fill the Crate	Turning Over a New Leaf Helium Hoop Saving Springfield Reflection Activities

Purpose List (continued) 

Purpose	Activity		
Peer Pressure/ Gossip and Rumors	US List Telegraph and Probability	Telegraph	"Follow Me"
Persistence/Focus	Play Dough Pictionary Charades Race Bag of Nouns Zoom Layers of the Atmosphere Telegraph and Probability	Collaborative List Making Telegraph Back Art Back-to-Back Drawing Helium Hoop Transition/Filler Activities	River Crossing Fill the Crate Challenge Turnstile Turning Over a New Leaf Helium Hoop Saving Springfield
Problem-Solving	Daily Riddle/Brainteaser Group Sculpture Dominos/Match Ups Zoom Layers of the Atmosphere Dominos Fraction Line Up	River Crossing Challenge Active Coordinate Plane Tin Can Pass Community Circles Zoom, Fill the Crate Turnstile	Turning Over a New Leaf Newspaper Bridges Helium Hoop Saving Springfield Time Management Reflection Activities
Reflection	Images/Postcards Toolbox/Metaphors Quotes Conversation Buttons Computer Keyboard Keys Journaling/Writing Prompts Question of the Day Group Sculpture Graffiti Wall Concentric Circles	Handshake Mingle Trade and Share Dominos/Match Ups Commonalities Mingle Anyone Who Pass the Knot Review/Reflection Dice Gallery Walk Play Dough Pictionary Charades Race	Layers of the Atmosphere Colleen's Lesson Team Tally Have You Ever? Celebration/Appreciation Personal Strengths Activities Group Norms Activities Time Management Activities Reflection Activities
Review and Application of Academic Content	Journaling/Writing Prompts Group Sculpture Concentric Circles Handshake Mingle Trade and Share Dominos/Match Ups Anyone Who Classroom Four Square Pass the Knot	Review/Reflection Dice Gallery Walk Play Dough Pictionary Charades Race Bag of Nouns Layers of the Atmosphere Dominos Fraction Line Up Telegraph and Probability Create a Colony	Colleen's Lesson Food Web Tag Active Coordinate Plane Collaborative List Making Team Tally Have You Ever? Reflection Activities
Self-Regulation/ Impulsivity	US List Classroom Four Square Play Dough Pictionary Charades Race Bag of Nouns	Telegraph and Probability Food Web Tag Collaborative List Making Have You Ever? Pass It To	Telegraph Time Management Transition/Filler Activities
Strengths Recognition/ Appreciation	Commonalities Mingle Name Meanings Peek-a-Who Names	Peek-A-Who Celebrations Celebration/Appreciation Personal Strengths Activities	Group Norms Activities Reflection Activities
Trust	US List Telegraph and Probability Name Meanings Tin Can Pass	Telegraph Communication Breakdown Turnstile, Turnstile	Turning Over a New Leaf Celebration/Appreciation Reflection Activities

Resources

Chapter One

Resources on the Brain and Learning

Books

Medina, John. (2008). *Brain Rules: 12 Principles for Surviving and Thriving at Work, Home and School.* Seattle, WA: Pear Press.

Sousa, D. A. (2010). *Mind, Brain, and Education: Neuroscience Implications for the Classroom.* Bloomington, IN: Solution Tree Press.

Willis, Judy. (2006). *Research-Based Strategies to Ignite Student Learning.* Alexandria, VA: ASCD.

Willis, Judy. (2008). *How Your Child Learns Best.* Naperville, IL: Sourcebooks.

Aamodt, S., & Wang, S. (2011). *Welcome to Your Child's Brain: How the Mind Grows from Conception to College.* New York: Bloomsbury.

Online Resources and Professional Development Organizations

The **Learning & the Brain Society** brings neuroscientists and educators together to explore new research on the brain and learning and its implications for education. Research centers and labs at Harvard, Yale, MIT, Stanford, UC BerkelyUniversity of Chicago, Johns Hopkins, and other leading institutes have joined forces with Learning & the Brain Society to provide the latest findings and co-sponsor professional development efforts • learningandthebrain.com

Dr. Judy Willis' website radteach.com offers links to her informative articles in magazines such as *Edutopia, Scholastic, Psychology Today* and more.

Dr. Sam Wang and Sandra Aamodt's website features facts, myths, and useful tips about your brain. This site offers an informative blog and helpful links to explore emerging information about the brain and learning • welcometoyourbrain.com

The Inspired Educator Blog by Jen Stanchfield offers brain-based teaching tips, strategies and resources • experientialtools.com/blog

Chapter Two

Resources on Differentiated Instruction

Books

Wormeli, R. (2007). *Differentiation: From Planning to Practice, Grades 6-12.* Portland, ME: Stenhouse Publishers.

Tomlinson, C. A. & Imbeau, M. (2010). *Leading and Managing a Differentiated Classroom.* Alexandria, VA: ASCD publisher.

Sousa, D., & Tomlinson, C. A. (2010). *Differentiation and Brain: How Neuroscience Supports the Learner-Friendly Classroom.* Bloomington, IN: Solution Tree Press

Online Resources

Dr. Carol Ann Tomlinson's website with links to presentations and resources • Caroltomlinson.com

21st Century Skills and Career Readiness Resources

Books

Trilling, B. & Fadel, C. (2009). *21st Century Skills: Learning for Life in Our Times.* San Francisco: Jossey-Bass/Wiley.

Wagner, Tony. (2012). *Creating Innovators: The Making of Young People Who Will Change the World.* New York: Scribner/Simon & Schuster.

Wagner, Tony. (2010). *The Global Achievement Gap: Why Even Our Best Schools Don't Teach the New Survival Skills Our Children Need—and What We Can Do About It.* New York: Basic Books.

Online Resources

The Partnership for 21st Century Skills is a national organization • p21.org

The Institute for Habits of Mind website explores the framework created by Art Costa and Bena Kallick. This framework offers 16 skill sets, attitudes, and qualities that will help learners succeed in the 21st Century • instituteforhabitsofmind.com

The 21st Century Fluency Project cultivates 21st Century fluencies while fostering engagement and adventure in the learning experience • fluency21.com

Resources for Social-Emotional Learning

Articles

Vega, Vanessa. (2014). *Social and Emotional Learning Research Review.* Edutopia Blog • edutopia.org/sel-research-learning-outcomes

Durlak, J., Weissberg, R. P., Dymnicki, A. B., Taylor, R. D., & Schellinger, K. B. (2011). The Impact of Enhancing Students' Social and Emotional Learning: A Meta-Analysis of School-Based Universal Interventions (PDF). *Child Development, 82(1),* 405-432.

Online Resources

In 2005, the state of Illinois developed social-emotional learning goals, standards and benchmarks for social-emotional learning in their public schools • isbe.state.il.us/ils/social_emotional/standards.htm

Strengths-Based Teaching and Facilitation Resources

Books

Duckworth, A.L., Peterson, C., Matthews, M.D., & Kelly, D.R. (2007). Grit: Perseverance and Passion for Long-Term Goals. *Personality Processes and Individual Differences, 92 (6),* p. 1087.

Dweck, C. S. (2006). *Mindset: The New Psychology of Success.* New York: Random House.

Pink, D. H. (2009). *Drive: The Surprising Truth About What Motivates Us.* New York: Riverhead Books.

Organizations/Online Resources

The Duckworth Lab at University of Pennsylvania focuses on researching the traits of self-control and grit to determine how they might predict both academic and professional success • https://sites.sas.upenn.edu/duckworth

The Search Institute has identified 40 concrete, positive experiences and qualities—developmental assets—that have a tremendous influence on young people's lives and choices • www.search-institute.org

The Original (now out of print) New Games Foundation's Books:
The New Games Book, 1976, edited by Andrew Fleugelman and published by Dolphin Books
More New Games!, 1981, edited by Andrew Fleugelman and published by Dolphin Books

Online Articles About the History of the New Games Movement:
Yehuda Blog Post from February 14th 2008: The History of the New Games Foundation: Play Hard, Play Fair. Nobody Hurt • http://jergames.blogspot.com/2008/02/history-of-new-games-foundation.html

Kristen Baumlier's Blog Post December 2011: Play Hard, Play Fair Nobody Hurt Overview of "New Games" • http://kristenbaumlier.com/2011/12/30/play-hard-play-fair-nobody-hurt-%E2%80%93-overview-of-%E2%80%9Cnew-games%E2%80%9D/

Chapter Four
Classroom Design

Erin Klein's Kleinspiration Teaching Blog: Kleinspiration.com

Standing Desks

Stand Up for Learning, Abby Brown • http://marine.stillwater.k12.mn.us

Safco products manufacturer of standing desks • safcoproducts.com

Chapter Five
"Hook" and Reflection Tools and Resources

Miniature Metaphors • experientialtools.com or contact jen@experientialtools.com

Pick-A-Postcard Kit • experientialtools.com or contact jen@experientialtools.com

Chiji Cards • The Institute for Experiential Education www.chiji.com

Conversation Starter Pin Back Buttons • experientialtools.com

Quotables • experientialtools.com or contact jen@experientialtools.com

Chat Packs Question Prompts • Questionguys.com

Table Topics Conversation Prompts • Tabletopics.com

Brain Teaser/Daily Riddle Ideas: *Mind Trap Game* • outsetmedia.com

The Me I See: Learning Through Writing and Reflection 2 Edition Reflective Prompts: Wood N Barnes Collective (2009).

Chapter Six

Chiji Processing Dice • www.chiji.com

Dry Erase Dice • learningadvantage.com • 866-564-8251

Chapter Seven/Eight

Activity Resources

Scattergories (List Making Review p.) • Hasbro.com

Battleship (Active Coordinate Plane) Battleship known worldwide as a pencil and paper strategy game dates from World War I. It was published by various companies as a pad-and-pencil game in the 1930s, and was released as a plastic board game by Milton Bradley Company in 1967. There are many variations found online.

Wilderdom.com • games and activity resources

Inspired Educator Blog at experientialtools.com offers activity ideas and articles on reflection, experiential education and brain-based learning

Fundoing.com • Chris Cavert's website

Zoom Banyai, I. (1995). *Zoom*. New York: Viking.

Community Puzzle • Communitypuzzle.com • training-wheels.com

Chapter Nine (See resources for Chapter Five and Six)

Reflection Tools and Resources

Thumballs • training-wheels.com

Spinner • mayer-johnson.com

Shapiro, Fred R. (2006). *The Yale Book of Quotations*. New, Haven, CT: Yale University Press.

Jane Jenni Personalities Pin Back Buttons • Janejenni.com or call 651.224-6763

Bickle, P., Carlin, C., Germanson, M. A., Gonzalez, J., & Huston, M. (2012). *The Bookmaking Experience: An Educator's Guide to Student Made Books*. Bethany, OK: Wood N Barnes Pubublishing.

Rapparlie, L. (2011). *Writing and Experiential Education: Practical Activities and Lesson Plans to Enrich Learning*. Bethany, OK: Wood N Barnes Pubublishing

Eckert, L. (2009). *If Anybody Asks Me...: 1,001 Questions for Educators, Counselors, and Therapists*. Bethany, OK: Wood N Barnes Pubublishing

Mindfulness Resources

Mindful Schools • mindfulschools.org

Mindfulness in Schools Project • Mindfulnessinschools.org

MindUP/The Hawn Foundation • thehawnfoundation.org

Organizations Highlighted Throughout This Book

LOOP New Orleans

Louisiana Outdoors Outreach Program-New Orleans (LOOP-New Orleans) provides outdoor educational and recreational programs for students in the greater New Orleans area. LOOP-New Orleans uses experiential techniques, environmental and adventure education and community building activities to enhance physical and social-emotional skills development and academic skills. Louisiana Outdoors Outreach Program LA Office of State Parks • (504) 388-7468 • crt.state.la.us/louisiana-state-parks/louisiana-outdoors-outreach-program/loop-new-orleans/index #1

Team BRIDGES - BRIDGES Center Memphis, TN
Team BRIDGES offers corporate and organizational customized team building, adventure education, leadership trainings and positive experiential school climate • programs.teambridges.org

Mentoring Partnerships of Long Island and New York
The mission of the Mentoring Partnership of Long Island is to bring caring adults together with children in need through safe, effective mentoring programs • mentorkids.org • 631-761-7800

The Experiment in International Living and Youth Development Team
Programs of World Learning • Extraordinary High School Summer Abroad Programs • 802 258-3444 The Experiment in International Living provides summer programs for high school students who want to connect deeply and engage meaningfully with the richness and complexities of another country. www.experimentinternational.org • youthprograms.worldlearning.org • www.worldlearning.org

Shaver's Creek Environmental Education and Team Development Center
The programs Shaver's Creek, Environmental Education Center, Outdoor School, Raptor Center, and Team Development Center provide a mix of educational and recreational opportunities for families, schools, corporate groups, and Penn State students. Pennsylvania State University • (814) 863 2000 • www.ShaversCreek.org/teams

Camp Runoia
Camp Runoia is a private overnight summer camp located in Belgrade Lakes, Maine that has been providing camp experiences for girls ages 7 to 16 since 1907. Camp Runoia's programs focus on personal growth and increased independence balanced with learning to be a part of a larger community • www.runoia.com • 207-495-2228

Schools Out Afterschool Program South Burlington School District Burlington, VT
South Burlington School District's "School's Out " program provides children a safe environment to express themselves and develop socially, physically, artistically, and creatively through a program that nurtures and respects the uniqueness of each and every child • acommunityofafterschool.blogspot.com

Twin Valley Middle-High School, 4299 VT RT 100, Whitingham, Vermont
Serving students in the southern Vermont towns of Whitingham, Jacksonville, Wilmington, Readsboro, Halifax, Searsburg, Marlboro, and West Dover, Twin Valley Middle-High School is a student-centered learning environment which respects and encourages the unique gifts within each individual. TVMS-TVHS seeks to foster scholarship, creativity, self-discipline, independent thinking, accountability, and respect. They strive to provide students with the knowledge, skills, and character to achieve personal success. Principal Tom Fitzgerald, Vice Principal Lee Ann Monroe • www.tvhas.k12.vt.us

Bibliography

Aamodt, S., & Wang, S. (2011). *Welcome to your child's brain: How the mind grows from conception to college.* New York: Bloomsbury.

Allday, R. A., & Pakurar, K. (2007). Effects of teacher greetings on student behavior. *Journal of Applied Behavioral Analysis, 40(2),* 317-320.

Anderson, G. L., Herr, K., & Nihlen, A. (2007). *Studying your own school: An educator's guide to practitioner action research (2nd edition).* Thousand Oaks, CA: Corwin Press.

Banyai, I. (1995). *Zoom.* New York: Viking.

Barker, J. E., Semenov, A. D., Michaelson, L., Provan, L. S., Snyder, H. R., & Munakata, Y. (2014). Less-structured time in children's daily lives predicts self-directed executive functioning. *Front. Psychol. 5:593.* Retrieved July 7, 2014 from: http://journal.frontiersin.org/Journal/10.3389/fpsyg.2014.00593/abstract.

Bartlett, T. (2014, May 7th). The Chronicle of Higher Education Blog. Retrieved from: http://chronicle.com/blogs/percolator/a-walk-in-the-park/34233#disqus_thread May 29th, 2014.

Bass. J., & Rankin, A. (1977). The Hobbit [Animated Movie]. New York: Rankin Bass Studio.

Batt, T. (2010). Using play to teach writing. *The American Journal of Play, 3(1),* 63-77.

Baumeister, R. F., & Bushman, B. J. (2008). *Social psychology and human nature.* Belmont, CA: Thomson Higher Education.

Benden, M. E., Wendel, M. L., Jeffrey, C. E., Hongwei, Z., & Morales, M. L. (2012). Within-subjects analysis of the effects of a stand-biased classroom intervention on energy expenditure. *Journal of Exercise Physiology Online, 15(2).*

Bickle, P., Carlin, C., Germanson, M. A., Gonzalez, J., & Huston, M. (2012). *The bookmaking experience: an educator's guide to student made books.* Bethany, OK: Wood N Barnes Publishing.

Blake, J. J., Benden, M. E., & Wendel, M. L. (2012) Using stand/sit workstations in classrooms: lessons learned from a pilot study in Texas. *Journal of Public Health Management Practice, Sep-Oct; 18(5)*: 412-5. Retrieved December 1, 2103, from http://www.ncbi.nlm.nih.gov/pubmed/22836531.

Bodrova, E., & Leong, D. J. (2007). *Tools of the mind* (Second ed.). Upper Saddle River, NJ: Pearson.

Borton, T. (1970). *Reach, touch, and teach; student concerns and process education.* New York: McGraw-Hill.

Bowman, S. L. (2005). *The ten-minute trainer: 150 ways to teach it quick and make it stick!* San Francisco: Pfeiffer.

Bransford, J. (2000). *How people learn: brain, mind, experience, and school* (Expanded ed.). Washington, DC: National Academy Press.

Bronson, P., & Merryman, A. (2010, July 10th) "The Creativity Crisis." *Newsweek.*

Bunce, D. M., Flens, E. A., & Neiles, K. Y. (2010). *How long can students pay attention in class? A study of student attention decline using clickers.* Journal of Chemical Education, 87(12), 1438-1443.

Cain, J., Cummings, M., & Stanchfield, J. (2005). *A teachable moment: A facilitator's guide to activities for processing, debriefing, reviewing and reflection.* Dubuque, IA: Kendall Hunt Pub.

Cain, J. H., & Jolliff, B. (1998). *Teamwork & teamplay: A guide to cooperative, challenge, and adventure activities.* Dubuque, IA: Kendall Hunt Pub.

Cain, J. H., & Smith, T. E. (2007). *The revised & expanded book of raccoon circles: A facilitator's guide to building unity, community, connection and teamwork through active learning.* Dubuque, IA: Kendall Hunt.

Calloway, C. (1939). *The new Cab Calloway's cat-ologue* [Rev. ed.]. S.l.: s.n.]. Retrieved October 10, 2013 from: http://www.tcswing.com/PDFs/Hepsters%20Dictionary.pdf

Carson, S. (2010). *Your creative brain: Seven steps to maximize imagination, productivity, and innovation in your life.* San Francisco: Jossey-Bass.

Carson, S. (2011, Nov.). "Creative Brains: Maximizing Imagination and Innovation in Students." Lecture Conducted at the Learning & the Brain Conference: Preparing 21st Century Minds. Boston, MA.

Cavert, C., & Frank, L. S. (1999). *Games (& other stuff) for teachers: Classroom activities that promote pro-social learning.* Bethany, OK: Wood N Barnes Publishing.

Cavert, C., & Simpson, S. (2010). *The Chiji guidebook: A collection of experiential activities and ideas for using Chiji cards.* Bethany, OK: Wood N Barnes Publishing.

Chapman, S. B. (2013, Nov 12th). Center for Brain Healthy Blog: Aerobic Exercise Improves Memory, Brain Function and Physical Fitness. Retrieved from http://www.brainhealth.utdallas.edu/blog_page/study-finds-aerobic-exercise-improves-memory-brain-function-and-physical-fitness June, 2014

Chapman, S. B., Aslan, S., Spence, J. S., DeFina, L. F., Keebler, M. W., Didehbani, N., & Lu, H. (2013). Shorter term aerobic exercise improves brain, cognition, and cardiovascular fitness in aging. *Frontiers in Aging Neuroscience. 5:75.* Retrieved fromhttp://journal.frontiersin.org/Journal/10.3389/fnagi.2013.00075/abstract

Cornford, F. M. (1934). *Plato's theory of knowledge: The Theatetus and the Sophist of Plato.* London: Routledge & Kegan Paul.

Csikszentmihalyi, M. (1990). *Flow: The psychology of optimal experience.* New York: Harper & Row.

Deardorff, J. (2012). Standing desks: The classroom of the future? *Chicago Tribune. August 7 edition* (online). Retrieved December 6, 2013, from http://articles.chicagotribune.com/2012-08-07/health/chi-standing-desks-the-classroom-of-the-future-20120807_1_desks-standings-classroom

Deluzain, E. (1996). *Names and personal identity.* Retrieved October 20th, 2013 from www.behindthename.com/articles/3.php.

Dewey, J. (1897). My Pedagogic Creed. *School Journal, 54,* 77-80. Found in McDermott, J. (1981), *The Philosophy of John Dewey: Two Volumes in One: The Structured Experience and The Lived Experience,* pp 442-454. Chicago: University of Chicago Press.

Dewey, J. (1902). *The child and the curriculum.* Chicago, University of Chicago Press. Retrieved from http://books.google.com/books

Dewey, J. (1900) *The school and society.* Chicago: The University of Chicago Press. Rev. Ed. (1915)

Dewey, J. (1910). *How we think.* Lexington, MA: D.C. Heath.

Dewey, J. (1916). *Democracy and Education. An introduction to the philosophy of education.* New York: Macmillan Co.

Dewey, J. (1925) (1959). Experience and Nature 2nd Ed., Chap. 1. And Chapter 10. Found in McDermott, J. (1981), *The Philosophy of John Dewey: Two Volumes in One: The Structured Experience and The Lived Experience.* Chicago: University of Chicago Press

Dewey, J. (1933). *How we think, a restatement of the relation of reflective thinking to the educative process.* Boston: D.C. Heath and Co.

Dewey, J. (1934). Art as Experience, pp 3-57. New York: Capricorn Books, 1958 (1934). Found in McDermott, J. (1981), *The Philosophy of John Dewey: Two Volumes in One: The Structured Experience and The Lived Experience.* Chicago: University of Chicago Press.

Dewey, J. (1938). *Experience and education.* New York: Macmillan.

Diaconis, P., Holmes, S., & Montgomery, R. (2007). Dynamical bias in the coin toss. *SIAM Review, 49(2),* 211-235.

Diamond, A. (2010). The evidence base for improving school outcomes by addressing the whole child and by addressing skills and attitudes, not just content. *Early Education and Development, 21(5),* 780-793. Retrieved February 10, 2013, from http://www.ncbi.nlm.nih.gov/pmc/articles/PMC3026344/

Di Stefano, G., Gino, F., Pisano, G. P., & Staats, B. R. (2014, March 25th). *Learning by Thinking: How Reflection Aids Performance.* Harvard Business School NOM Unit Working Paper No. 14-093; Harvard Business School Technology & Operations Mgt. Unit Working Paper No. 14-093. Available at SSRN: http://ssrn.com/abstract=2414478

Dixon, J. K., Egendoerfer, L. A., & Clements, T. (2009, Nov). Do they really need to raise their hands? Challenging a traditional social norm in a second grade mathematics classroom. *Teaching and Teacher Education: An International Journal of Research and Studies, 25:8,* p1067-1076.

Duckworth, A. L., Peterson, C., Matthews, M. D., & Kelly, D. R. (2007). Grit: Perseverance and passion for long-term goals. *Personality Processes and Individual Differences, 92 (6),* p. 1087.

Durlak, J., Weissberg, R. P., Dymnicki, A. B., Taylor, R. D., & Schellinger, K. B. (2011). The Impact of Enhancing Students' Social and Emotional Learning: A Meta-Analysis of School-Based Universal Interventions (PDF). Child Development, 82(1), 405-432.

Dweck, C. S. (2006). *Mindset: The new psychology of success.* New York: Random House.

Dweck, C. (2006). Mindset: *The new psychology of success.* New York: Ballantine Books.

Fadel, C. (2012, Nov). "The Neuroscience of Learning Educating Diverse Minds." The Learning and the Brain Society Conference Keynote Lecture. Boston, MA.

Fischer, K., Goswami, U., Geake, J., & the Task Force on the Future of Educational Neuroscience. (2010). The Future of Educational Neuroscience. *Mind, Brain and Education, 4:2,* pages 68-80.

Fluegelman, A. (1976). *The new games book.* Garden City, NY: Dolphin Books.

Fluegelman, A. (1981). *More new games!—and playful ideas from the New Games Foundation.* New York: Dolphin Books/Doubleday.

Gass, M.A. (1995). *Book of metaphors: Volume II.* Dubuque, IA: Kendall Hunt.

Goswami. U. (2004). Neuroscience and education. *British Journal of Educational Psychology, 74,* 1-14.

Gregoire, C. (2013, March). "Mindfulness programs in schools reduce symptoms of depression among adolescents: Study." The Huffington Post. Retrieved from: http://www.huffingtonpost.com/2013/03/15/mindfulness-in-schools-re_n_2884436.html November 5, 2013

Hillenburg, S. (1999). *SpongeBob Square Pants* [TV Series]. New York: Nickelodeon.

Illinois State Board of Education. (2005). Illinois learning standards: Social and emotional learning. Retrieved from http://www.isbe.net/ils/social_emotional/standards.htm

Immordino-Yang, M. H. (2012, April). "The Neuroscience of Social Emotion & Its Importance to Learning." The Learning and the Brain Society. Emotions and the Brain Symposium. New York.

Immordino-Yang, M. H. & Damasio A. R. (2007). We feel, therefore we learn: The relevance of affective and social neuroscience to education. *Mind, Brain, and Education, 1(1),* 3-10.

Immordino-Yang, M. H., & Faeth, M. (2010). The role of emotion and skilled intuition in learning. In D. A. Sousa (Ed), *Mind, brain, and education: Neuroscience implications for the classroom* (pp. 68-83). Bloomington, IN: Solution Tree Press.

Index to Group Activities, Games, Exercises & Initiatives. (n.d.). Index to Group Activities, Games, Exercises & Initiatives. Retrieved December 18, 2012, from http://wilderdom.com/games/

Jackson, T. (1993). *Activities that teach.* Cedar City, UT: Red Rock Pub.

Jackson, T. (1995). *More activities that teach.* Cedar City, UT: Red Rock Pub.

Jones, A. (1998). *104 activities that build: self-esteem, teamwork, communication, anger management, self-discovery, and coping skills.* Richland, WA: Rec Room Pub.

Kandel, E. R. (2006). *In search of memory: the emergence of a new science of mind.* New York: W. W. Norton & Co.

Kehret, P. (1996). *Small steps: The year I got polio.* Park Ridge, IL: Albert Whitman Co.

Kestenbaum, D. (Science Correspondent). (2004, February, 24). The not so random coin toss. *All Things Considered.* Washington, DC: National Public Radio.

Kilbourne, J. (2009). Sharpening the mind through movement: Using exercise balls in a university lecture class. *The Chronicle of Kinesiology and Physical Education in Higher Education. February 2009.*

Kilbourne, J. (2013). *Moving physical education beyond the gymnasium: Activity permissible classrooms.* Retrieved November 8, 2013 from http://www.pelinks4u.org/articles/kilbourne5_2013.htm.

Krown, M. K. (2009). *What is Vision Quest and why do one?* Huffington Post Blog 6/18/09 Retrieved November 6, 2013. from http://www.huffingtonpost.com/maddisen-k-krown/ask-maddisen-what-is-a-vi_b_217432.html.

Lacey, S., Stilla, R., & Sathian, K. (2012). Metaphorically feeling: Comprehending textural metaphors activates Somatosensory Cortex. *Brain Language, 120(3),* and 416-421. Retrieved August 8, 2013, from http://www.ncbi.nlm.nih.gov/pmc/articles/PMC3318916/.

Lowry, L. (1993). *The giver.* New York: Houghton Mifflin Harcourt.

McDermott, J. (1981) *The philosophy of John Dewey (two volumes in one): 1. The structured experience; 2. The lived experience.* Chicago: University of Chicago Press.

Medina, J. (2008). *Brain rules: 12 principles for surviving and thriving at work, home, and school.* Seattle, WA: Pear Press.

Neighmond, P. (Health Policy Correspondent). (August 31, 2006). *Exercise helps students in the classroom* [Radio series episode]. In Morning Edition. Boston: NPR.

O'Hanlon, L. (2013). *To foster productivity and creativity in class, ditch the desks!* Mindshift Education Blog by NPR Station KQED Retrieved from http://blogs.kqed.org/mindshift/2013/08/to-foster-productivity-and-creativity-in-class-ditch-the-desks/ October 2, 2013

Oppezzo, M., & Schwartz, D. L. (2014, April 21). Give Your Ideas Some Legs: The Positive Effect of Walking on Creative Thinking. *Journal of Experimental Psychology: Learning, Memory, and Cognition.* Advance online publication. http://dx.doi.org/10.1037/a003657

Ormrod, J. E. (2012). *Essentials of educational psychology: Big ideas to guide effective teaching.* Boston, MA: Pearson Education Inc.

Oz, F. (Director). (1991). *What about Bob?* [Motion picture]. U.S.: Continental film.

Piaget, J. (1926). *The language and thought of the child.* London: Routledge & Kegan.

Piaget, J. (1953). *The origin of intelligence in the child.* New York: Routledge & Kegan.

Piaget, J. (1976). *Piaget sampler: An introduction to Jean Piaget through his own words.* New York: Wiley.

Pink, D. H. (2009). *Drive: the surprising truth about what motivates us.* New York: Riverhead Books.

Rapparlie, L. (2011). *Writing and experiential education: practical activities and lesson plans to enrich learning.* Bethany, OK: Wood N Barnes Publishing.

Ratey, J. J., & Hagerman, E. (2008). *Spark: the revolutionary new science of exercise and the brain.* New York: Little, Brown.

Rath, T., & Buckingham, M. (2007). *StrengthsFinder 2.0* (New & upgraded ed.). New York: Gallup Press.

Reiff, C. J., Marlatt, K. L., & Dengel, D. R. (2012). Difference in caloric expenditure in sitting versus standing desks. *Journal of Physical Activity and Health, 9,* 1009-1011.

Rogers, C. R. (1969). *Freedom to Learn.* Columbus, OH: Merrill.

Rogers, C. R. (1983). *Freedom to learn for the 80's.* Columbus, OH: C.E. Merrill Pub. Co.

Rohnke, K. (1994). *The bottomless bag again (2nd ed.).* Dubuque, IA: Kendall Hunt.

Rohnke, K. (2002). *A small book about large group games.* Dubuque, IA: Kendall Hunt.

Rohnke, K. (2004). *Funn 'n games.* Dubuque, IA: Kendall Hunt.

Rohnke, K. (2009). *Silver bullets: A revised guide to initiative problems, adventure games, and trust activities.* Dubuque, IA: Kendall Hunt.

Rohnke, K., & Butler, S. (1995). *Quicksilver: Adventure games, initiative problems, trust activities, and a guide to effective leadership.* Dubuque, IA: Kendall Hunt.

Schon, D. A. (1983). *The reflective practitioner: How professionals think in action.* New York: Basic Books, Inc.

Sikes, S. (1998). *Executive marbles and other team-building activities.* Richmond, TX: Learning Unlimited Inc.

Simpson, S. (2003). *The leader who is hardly known: Self-less teaching from the Chinese tradition.* Bethany, OK: Wood N Barnes.

Sousa, D. A. (2006). *How the brain learns (3rd ed.).* Thousand Oaks, Calif.: Corwin Press.

Sousa, D. A. (2010). *Mind, brain, and education: Neuroscience implications for the classroom.* Bloomington, IN: Solution Tree Press.

Sousa, D. (2012, Nov.). "Differentiation and the Brain: How Neuroscience Supports the Learner-Friendly Classroom." Lecture conducted at The Learning & the Brain Conference: Educating Diverse Minds. Boston, MA.

Stanchfield, J. (2001). Adventures in Learning: Methods and Materials for Aiding Teachers in the Implementation of Experiential Education Curriculum in Mankato, MN. APP Paper. Minnesota State University, Mankato pp 14-15.

Stanchfield, J. (2002). *Middleton High School Adventure Education Curriculum.*

Stanchfield, J. (2007). *Tips & tools: The art of experiential group facilitation.* Bethany, OK: Wood N Barnes Publishig.

Stanchfield, J. (2010). *Processing in the middle of the experience.* The Inspired Educator Blog, Retrieved from http: http://www.experientialtools.com/blog.

Stanchfield, J. (2011). Bookending a learning experience with stong beginnings and endings. Another fun idea: Upcycled computer key board keys. The Inspired Educator Blog, Retrieved from www.experientialtools.com/blog.

Tolkien, J. R. R. (1937). *The Hobbit*, or, *There and back again*. London: George Allen & Unwin.

Tomlinson, C. A. (2012, Nov). "Differentiated Classroom: Five Elements That Make Classrooms Work for All Students." Lecture Conducted at the Learning & the Brain Conference: Educating Diverse Minds. Boston, MA.

Tomlinson, C. A. (2013, April). "The Brain and Differentiation." Keynote Luncheon Presentation. New England League of Middle Schools Conference. Providence, RI.

Trilling, B. & Fadel, C. (2009). *21st century skills: Learning for life in our times.* San Francisco: Jossey-Bass/Wiley.

Vega, Vanessa (2014) Social and Emotional Learning Research Review. Edutopia Blog Retrieved http://www.edutopia.org/sel-research-learning-outcomes.

Vescio, V., Ross, D., & Adams, A. (2008). A review of research on the impact of professional learning communities on teaching practice and student learning. *Teaching and Teacher Education 24.* Authors Personal Copy Retrieved Online Retrieved May 28, 2014https://www.k12.wa.us/Compensation/pubdocs/Vescio2008PLC-paper.pdf.

Vygotsky, L. (1933). Play and its role in the mental development of the child. Psychology and Marxism Internet Archive: Lev Vygotsky Archive, Voprosy psikhologii, 1966, No. 6; (Translated by Catherine Mulholland). Retrieved November 15, 2012, from http://www.marxists.org/archive/vygotsky/works/1933/play.htm.

Vygotsky, L. S., & Cole, M. (1978). *Mind in society: The development of higher psychological processes.* Cambridge: Harvard University Press.

Wagner, T. (2011, Nov). "Teaching, Learning and Leading in the 21st Century." Keynote Presentation. Learning & the Brain Conference: Preparing 21st Century Minds. Boston, MA.
Wagner, T. (2012). *Creating innovators: The making of young people who will change the world.* New York: Scribner.
Wang, S. (2014, April 19th). Personal conversation.
Weinstein, L., Laverghetta, A., Alexander, R., & Stewart, M. (2009). Teacher greetings increase college students' test scores. *College Student Journal, 43(2)*, 452-453.
Weinstein, M., & Goodman, J. B. (1985). *Everybody's guide to noncompetitive play: Playfair.* San Luis Obispo, CA: Impact Publishers.
Whirty, Ryan (2002). Ordinary Genius. *Research & Creativity Activity. Indiana University XXV: 1.* Retrieved November 2, 2013 from http://www.indiana.edu/~rcapub/v25n1/genius.shtml.
Whitehead, A. N. (1929). *The aims of education & other essays.* New York: Macmillan Co..
Wiliam, D. (2011) "Embedded Formative Assessment." Presentation at the AuthorSpeak Conference, Indianapolis, IN. November 1, 2011. [PowerPoint Slides] Retrieved April 9, 2014 from: www.dylanwiliam.org/Dylan_Wiliams_website/Presentations_files/Authorspeak_rev.pptx
Willis, J. (2006). *Research-based strategies to ignite student learning insights from a neurologist and classroom teacher.* Alexandria, VA: Association for Supervision and Curriculum Development.
Willis, J. (2007). *Brain-friendly strategies for the inclusion classroom insights from a neurologist and classroom teacher.* Alexandria, VA: Association for Supervision and Curriculum Development.
Willis, J. (2008). *How your child learns best: Brain-friendly strategies you can use to ignite your child's learning and increase school success.* Naperville, IL: Sourcebooks.
Willis, J. (2010a). The current impact of neuroscience on teaching and learning. In D. A. Sousa (Ed), *Mind, brain, and education: Neuroscience implications for the classroom* (pp. 44-66). Bloomington, IN: Solution Tree Press.
Willis, J. (2010b). *Understanding and planning achievable challenge. Learning to love math teaching strategies that change student attitudes and get results* (pp. 16-32). Alexandria, VA: ASCD.
Willis, J. (2011a) A neurologist makes the case for the video game model as a learning tool. *Edutopia Blog.* April 14, 2011. Retrieved June 1st, 2011 from http://www.edutopia.org/blog/video-games-learning-student-engagement-judy-willis.
Willis, J. (2011b, Nov). "Instruction and Curriculum for 21st Century Brains." Lecture Conducted at the Learning & the Brain Conference: Preparing 21st Century Minds. The Learning and the Brain Society, Boston, MA.
Willis, J. (2012, Nov). "Neuroscience of the Adolescent Brain Reveals Challenges and Opportunities for Positive Interventions." Lecture conducted at The Learning & the Brain Conference: Educating Diverse Minds. Boston, MA.
Willis, J. (2013). *Unlocking student-directed learning and concept acquisition*. Lecture conducted at The Learning & the Brain Conference: Engaging 21st Century Minds. Boston, MA.
Wood N Barnes Collective. (2009). *The me I see: Learning through writing and reflection (2nd ed.).* Bethany, OK: Wood N Barnes Publishing.
Yancey, A. K. (2010). *Instant recess: Building a fit nation 10 minutes at a time.* Berkeley: University of California Press.
Zeichner, K., & Liston, D. (2013). *Reflective teaching (2nd edition).* New York: Routledge.
Zimmerman, J. L. (2013). *What (most) schools fail to teach.* The Huffington Post Blog .09/13/2013. Retrieved November 13, 2103 from: http://www.huffingtonpost.com/jamie-lauren-zimmerman/meditation-schools_b_3881544.html.

Index

Photo by Antje Kastner's Studio

Jennifer Stanchfield's depth of experience, creativity and knowledge of educational theory and practice is evident in her innovative yet practical workshops and publications in which she incorporates the art of facilitation and teaching with brain-based and pedagogical research. In her 25 years as an educator, Jen has worked as a classroom teacher, a clinician in mental health treatment centers for children, adolescents and adults, as an adventure educator, and in professional training, adult education, and organizational teambuilding. Jen earned her masters degree in Experiential Education at Minnesota State University, Mankato and continues to pursue the latest research on the brain and learning, pedagogy and the emerging field of educational neuroscience. Through these experiences she has developed an extensive repertoire of evidence-informed experiential activities, tools, and strategies she brings to her engaging and informative workshops, publications, and teaching resources. She works with schools, colleges, community organizations and training organizations all over the world—helping educators increase engagement, promote social-emotional skills and facilitate meaningful reflection and group dialogue to increase learning outcomes.

Jen is founder and director of Experiential Tools, an organization focused on helping educators fill their toolbox with quality, unique, and user-friendly methods to enhance learning. Her group facilitation and teaching tools include the Pick-A-Postcard Reflection Kit, the Miniature Metaphors Processing and Reflection Treasure Chest, Quotables, and the Conversation Starter Pin-Back Button collection. She is author of *Tips and Tools: The Art of Experiential Group Facilitation* and co-author of *A Teachable Moment: A Facilitator's Guide to Activities for Processing, Debriefing, Reviewing and Reflection.* She is also the creator and a regular contributor to the byline "Facilitators Toolbox" published by the Association for Challenge Course Technology (ACCT) and "The Inspired Educator Blog" at experientialtools.com.

Along with training educators and working directly with schools, Jen provides team-building and professional development programs for community organizations, businesses, college staff and faculty. She offers custom as well as open enrollment workshops for counselors, trainers, classroom teachers, college programs, coaches, and human resource professionals looking to enliven their curriculum or group work with experiential, brain-based approaches to increase engagement and create lasting lessons.

For more information on workshops, consulting services, The Inspired Educator Blog, and facilitation and teaching tools visit jenniferstanchfield.com or experientialtools.com.

Other books by Jennifer Stanchfield

Tips & Tools: The Art of Experiential Group Facilitation

A Teachable Moment:
A Facilitator's Guide to Activities for Processing, Debriefing, Reviewing and Reflection
Cain, Cummings & Stanchfield